Robert J. Nicholls · W. Neil Adger ·
Craig W. Hutton · Susan E. Hanson
Editors

Deltas in the Anthropocene

Editors
Robert J. Nicholls
School of Engineering
University of Southampton
Southampton, UK

W. Neil Adger
Geography, College of Life
and Environmental Sciences
University of Exeter
Exeter, UK

Craig W. Hutton
GeoData Institute, Geography
and Environmental Science
University of Southampton
Southampton, UK

Susan E. Hanson
School of Engineering
University of Southampton
Southampton, UK

https://doi.org/10.1007/978-3-030-23517-8

We dedicate this book to our DECCMA colleague Dr. Asish Kumar Ghosh, who passed away in April 2018 at the age of 80 years.

Dr. Ghosh founded the Centre for Environment and Development, Kolkata and was Director of the Zoological Survey of India. He served the Government of India for more than two decades under the Ministry of Environment and Forests, was a Fulbright scholar, and led the Indian delegation to the original Ramsar Convention in Japan. He was instrumental in reintroducing salt-tolerant rice varieties in the Indian Sundarbans after Cyclone Aila.

Dr. Ghosh had a lifelong commitment to the environment, biodiversity, and human rights in India and throughout South Asia and shared his great knowledge and wisdom with both students and colleagues, many of whom have contributed to this volume.

Foreword

This book provides a quality excursion into one of the hottest topics in environmental research. Deltas have historically offered up a rich potential in the maintenance of coastal biodiversity, and flow regulation between the landscape and the coastal ocean. Deltas remain a key environment for human development and support. The 600 million habitants, now living or working on deltas, face the looming threat of global sea-level rise due to climate change. More local coastal subsidence, related to water extraction (urban consumption, irrigation, and industrialisation), peat oxidation, and petroleum mining, further compounds the impact of a rising global ocean.

Deltas are key engines of the global economy, and the destination of the largest migration of humans in history. Once the world's rice bowls, many deltas have been transformed into protein bowls, with new aquaculture infrastructure increasing the ecosystem services that they can render. To accomplish this feat, deforestation in the coastal zone (global loss of mangrove forests) has often been the result. And with the removal of coastal forests, the extent of storm surge inundation has increased, putting ever more of delta-citizens at risk. Further, marine inundation increases the risk of salinisation leading to depletion of agricultural soil and freshwater marshes, and thus a reduction in biodiversity.

There is a direct link between how a particular delta functioned in 1950, at the beginning of the Anthropocene Epoch, and its subsequent transformation with a rapidly increasing population. Over the last 65 years, megacities (10+ million residents) have sprung up on deltas, worldwide, increasing the need to protect these cities behind artificial barriers, or face reverse migration off delta, or both. Already officials in Bangkok and Jakarta, have developed plans to consider such a costly endeavour.

Deltas in the Anthropocene brings together a very large cast of researchers, to deal with the major issues concerning: (1) the Ganga-Brahmaputra-Meghna Delta of Bangladesh and India, (2) the Mahanadi Delta of India, and (3) the Volta Delta of Ghana. Quite often, insights from studies on other deltas of the world are discussed for comparison and guidance when discussing policy implications (e.g., livelihoods, housing infrastructure, health, and the impacts of gender on migration). Analysis includes socio-economic investigations into the temporal trends and spatial heterogeneity of the key factors affecting each delta, including the application of integrated modelling to make sense of the complexity underpinning each environmental system. Often the authors take the position that future natural flows are being replaced by human-controlled flows, whether one tracks water, sediment, food production, and so on. This suggests that deltas in these three delta regions will continue their transition towards a human-controlled, or Anthropogenic, environment.

I grew up living on a delta. My memories include the flatness of their topography; the profusion of seasonal bugs and flies that lived around the swamps, marshes and delta lakes; floods that would so scare us that we would often rethink our habitat; the rapid and historic growth of infrastructure and other forms of delta taming; and, of course, sand and mud everywhere. Huge volumes of agricultural and forestry goods, minerals, and manufactured goods were always being offloaded or loaded onto ships. The poorest people lived in the worse parts of the delta; the richest people always seemed to find highland for views and security.

Later in life, as I travelled the world, often with the opportunity to visit or study the local delta(s), I continued to explore how deltas functioned. It became clear to me, that any investigation of a modern delta, without accounting for their past and future human modifications, would miss the "story" of the delta. I also realised how there is a never-ending tension between the uplanders (those that lived upstream of the delta), and the lowlanders, the delta inhabitants. Uplanders considered deltas simply as home to valuable ports to distribute their upland goods onto other parts of the world economy. Not much thought was ever given to the impact of upstream dam building, for example, on a delta's coastal environment. Trapping of sediment in upstream reservoirs, and the subsequent erosion on a delta's shoreline, were not part of the general conversation.

We can do better, and it is not too late to start having the very important discussion on the governance and function of our world's deltas. I am so pleased that Canada's International Development and Research Center (IDRC) teamed up with the UK's Department for International Development (DFID) to create the Collaborative Adaptation Research Initiative in Africa and Asia or CARIAA programme that underwrote some of the important funding to support the book's contents. As an advisor to the CARIAA programme, I watched the progress of key aspects of the book's content. I am extremely pleased with the results that have broadened my own ability to contextualise deltaic environments in the Anthropocene. What makes this particular book on deltas exceptional, is that the researchers are largely from Ghana, India, and Bangladesh. A unique perspective is thus provided.

I have known the book's lead Editor, Robert J. Nicholls, for many years, having followed his applied science approach to issues related to coastal environmental science. This effort will no doubt be another feather in his cap, but also for his co-editors Neil Adger, Craig W. Hutton, and Susan E. Hanson. They have done an exceptional job, as have all 60 contributors. For those wanting a heads up on modern environmental science, this text has much to offer.

Boulder, CO, USA

Jaia Syvitski
University of Colorado

Acknowledgements

This book is the culmination of a major interdisciplinary research collaboration across biophysical and social sciences focussed on deltas. The collaboration has been facilitated through a major consortium funded between 2014 and 2018: the "Deltas, vulnerability and Climate Change: Migration and Adaptation" (DECCMA) project (IDRC 107642) under the Collaborative Adaptation Research Initiative in Africa and Asia (CARIAA) programme with financial support from the UK Government's Department for International Development and the International Development Research Centre, Canada. We thank them for this funding and their support, including making this book an open access publication. Thanks also go to the European Union Delegation to India who provided additional support through "Climate Adaptation and Services Community" contract no. ICI+/2014/342-806 for the research discussed in Chapter 9. The Delta Dynamic Integrated Emulator Model (ΔDIEM) used in Chapter 10 was developed as part of ESPA Deltas (Assessing Health, Livelihoods, Ecosystem Services and Poverty Alleviation in Populous Deltas) NE-J002755-1, funded by the Ecosystem Services for Poverty Alleviation programme.

The DECCMA consortium is a team of 27 partners located in Ghana, India, Bangladesh, Spain, Italy, and South Africa, as well as the United Kingdom (see www.deccma.com). To all the participants we extend our appreciation of their contribution to the achievements of the project. In particular, we thank the national leads without whom the project would not have been successful: Samuel Nii Ardey Codjoe and Kwasi Appeaning Addo (University of Ghana), Sugata Hazara and Tuhin Ghosh (Jadavpur University), and Munsur Rahman and Mashfiqus Salehin (Bangladesh University of Engineering and Technology). We are also extremely grateful to Jon Lawn and Lucy Graves (University of Southampton), Gertrude Owusu (University of Ghana), Sumana Banerjee (Jadavpur University), and Md. Anisur Rahman Majumdar (Bangladesh University of Engineering and Technology) who kept the machine running with their management and coordination of the project and country teams over nearly five years. Ms. Lyn Ertl (University of Southampton) kindly created Figs. 1.3 and 6.2 for this book. Last but not least, we also thank our IDRC Project Officer, Dr. Michele Leone, for his constant support and encouragement throughout the life of DECCMA.

CARIAA
Collaborative Adaptation Research Initiative in Africa and Asia

IDRC | CRDI
International Development Research Centre
Centre de recherches pour le développement international
Canada

UKaid
from the British people

Contents

Contributors

Mumuni Abu Regional Institute for Population Studies, University of Ghana, Legon-Accra, Ghana

Cynthia Addoquaye Tagoe Institute of Statistical, Social and Economic Research, University of Ghana, Legon-Accra, Ghana

W. Neil Adger Geography, College of Life and Environmental Sciences, University of Exeter, Exeter, UK

Prince Osei-Wusu Adjei Department of Geography and Rural Development, Kwame Nkrumah University of Science and Technology, Kumasi, Ghana

Andrew Allan School of Law, University of Dundee, Dundee, UK

Barnabas Akurigo Amisigo Water Research Institute, Council for Scientific and Industrial Research, Accra, Ghana

Fiifi Amoako-Johnson University of Cape Coast, Cape Coast, Ghana

Kirk Anderson Regional Institute for Population Studies, University of Ghana, Legon-Accra, Ghana

Kwasi Appeaning Addo Department of Marine and Fisheries Sciences, Institute for Environment and Sanitation Studies, University of Ghana, Legon-Accra, Ghana

Iñaki Arto bc^3—Basque Centre for Climate Change, Bilbao, Spain

Joseph Kwadwo Asenso Ministry of Finance, Government of Ghana, Accra, Ghana

D. Yaw Atiglo Regional Institute for Population Studies, University of Ghana, Legon-Accra, Ghana

Jennifer Ayamga Institute for Environment and Sanitation Studies, University of Ghana, Legon-Accra, Ghana

Rabindra N. Bhattacharya Department of Economics, Jadavpur University, Kolkata, India

Mohammad Rashed Alam Bhuiyan Refugee and Migratory Movements Research Unit, University of Dhaka, Dhaka, Bangladesh

Ignacio Cazcarro bc^3—Basque Centre for Climate Change, Bilbao, Spain;
Department of Economic Analysis, Aragonese Agency for Research and Development, Agrifood Institute of Aragon, University of Zaragoza, Saragossa, Spain

Alexander Chapman School of Engineering, University of Southampton, Southampton, UK

Samuel Nii Ardey Codjoe Regional Institute for Population Studies, University of Ghana, Legon-Accra, Ghana

Stephen E. Darby Geography and Environmental Science, University of Southampton, Southampton, UK

Shouvik Das School of Oceanographic Studies, Jadavpur University, Kolkata, India

Sophie Day Geography and Environmental Science, University of Southampton, Southampton, UK

Frances Dunn GeoData Institute, Geography and Environmental Science, University of Southampton, Southampton, UK

Emmanuel Ekow Asmah Department of Economics, University of Cape Coast, Cape Coast, Ghana

Amit Ghosh School of Oceanographic Studies, Jadavpur University, Kolkata, India

Tuhin Ghosh School of Oceanographic Studies, Jadavpur University, Kolkata, India

Susan E. Hanson School of Engineering, University of Southampton, Southampton, UK

Anisul Haque Institute of Water and Flood Management, Bangladesh University of Engineering and Technology, Dhaka, Bangladesh

Somnath Hazra School of Oceanographic Studies, Jadavpur University, Kolkata, India

Sugata Hazra School of Oceanographic Studies, Jadavpur University, Kolkata, India

Chris Hill GeoData Institute, Geography and Environmental Science, University of Southampton, Southampton, UK

Mohammed Abed Hossain Institute of Water and Flood Management, Bangladesh University of Engineering and Technology, Dhaka, Bangladesh

Craig W. Hutton GeoData Institute, Geography and Environmental Science, University of Southampton, Southampton, UK

Nabiul Islam Bangladesh Institute of Development Studies, Dhaka, Bangladesh

Philip-Neri Jayson-Quashigah Department of Marine and Fisheries Sciences, Institute for Environment and Sanitation Studies, University of Ghana, Legon-Accra, Ghana

Abiy S. Kebede School of Engineering, University of Southampton, Southampton, UK

Attila N. Lázár School of Engineering, University of Southampton, Southampton, UK

Anil Markandya bc[3]—Basque Centre for Climate Change, Bilbao, Spain

Francisca Martey Research Department, Ghana Meteorological Agency, Legon-Accra, Ghana

Adelina Mensah Institute for Environment and Sanitation Studies, University of Ghana, Legon-Accra, Ghana

Colette Mortreux Geography, College of Life and Environmental Sciences, University of Exeter, Exeter, UK

Winfred A. Nelson National Development Planning Commission, Accra, Ghana

Robert J. Nicholls School of Engineering, University of Southampton, Southampton, UK

Benjamin Kofi Nyarko Geography and Regional Planning, University of Cape Coast, Cape Coast, Ghana

Patrick K. Ofori-Danson Department of Marine and Fisheries Sciences, Institute for Environment and Sanitation Studies, University of Ghana, Legon-Accra, Ghana

Gertrude Owusu Regional Institute for Population Studies, University of Ghana, Legon-Accra, Ghana

Amrita Patel Sansristi, Bhubaneswar, Odisha, India

Giorgia Prati Geography and Environmental Science, University of Southampton, Southampton, UK

Victoria Price Geography and Environmental Science, University of Southampton, Southampton, UK

Ruth Maku Quaye Regional Institute for Population Studies, University of Ghana, Legon-Accra, Ghana

Md. Munsur Rahman Institute of Water and Flood Management, Bangladesh University of Engineering and Technology, Dhaka, Bangladesh

Rezaur Rahman Institute of Water and Flood Management, Bangladesh University of Engineering and Technology, Dhaka, Bangladesh

Pokkuluri Venkat Raju National Remote Sensing Center, Indian Space Research Organisation, Hyderabad, India

Mahmudol Hasan Rocky Refugee and Migratory Movements Research Unit, University of Dhaka, Dhaka, Bangladesh

Ricardo Safra de Campos Geography, College of Life and Environmental Sciences, University of Exeter, Exeter, UK

Mashfiqus Salehin Institute of Water and Flood Management, Bangladesh University of Engineering and Technology, Dhaka, Bangladesh

Maminul Haque Sarker Center for Environmental and Geographic Information Services, Dhaka, Bangladesh

Tasneem Siddiqui Refugee and Migratory Movements Research Unit, University of Dhaka, Dhaka, Bangladesh

Natalie Suckall Geography and Environmental Science, University of Southampton, Southampton, UK

Emma L. Tompkins Geography and Environmental Science, University of Southampton, Southampton, UK

Katharine Vincent Kulima Integrated Development Solutions, Pietermaritzburg, South Africa

List of Figures

List of Tables

List of Boxes

1

Delta Challenges and Trade-Offs from the Holocene to the Anthropocene

Robert J. Nicholls, W. Neil Adger, Craig W. Hutton and Susan E. Hanson

1.1 Introduction

The human dominance of the Earth and its implications is now captured in the concept of a fundamental transition to the Anthropocene. The Anthropocene represents a period of time when humans are the dominant influence on the climate and environment, as opposed to earlier periods more dominated by natural processes (Steffen et al. 2011; Ribot 2014;

R. J. Nicholls (✉) · S. E. Hanson
School of Engineering, University of Southampton, Southampton, UK
e-mail: r.j.nicholls@soton.ac.uk

W. N. Adger
Geography, College of Life and Environmental Sciences,
University of Exeter, Exeter, UK

C. W. Hutton
GeoData Institute, Geography and Environmental Science,
University of Southampton, Southampton, UK

R. J. Nicholls et al. (eds.), *Deltas in the Anthropocene*,
https://doi.org/10.1007/978-3-030-23517-8_1

Goudie and Viles 2016; Verburg et al. 2016; Donges et al. 2017). This transition to human domination is increasingly placed in the mid-twentieth century, though subject to vigorous debate. In this book, the focus is what the Anthropocene means for environmental management and achievement of the widely agreed goals of sustainable development (Hutton et al. 2018; Szabo et al. 2018), using deltas as hotspots where natural processes and intense and growing human activity intersect (Renaud et al. 2013).

Many relevant processes of environmental, economic and social change are progressing faster and more intensely in deltas than their global averages (Table 1.1). As humans have occupied deltas for centuries and millennia, they are, in effect, a bell-weather for the earth and the Anthropocene. Figure 1.1 illustrates some key features of deltas in the Anthropocene.

Deltas have formed at the land–sea interface over hundreds and thousands of years where large rivers deposit their sediment load creating extensive highly productive and low-lying coastal plains (Ibáñez et al. 2019). Natural deltas represent the interplay of sediment delivery and reworking, destructive marine processes and subsidence, including major river channel migration and switching (Syvitski 2008). With their extensive ecosystem services and accessible transport links, deltas have also been a focus for human settlement for millennia (Bianchi 2016). Their populations have grown dramatically in the past 100 years and today

Table 1.1 Key trends across deltas and globally, indicating a greater intensity in populated deltas

Issue	Delta trend	Global trend
Population	^^	^^
Economy	^^	^
Cities and urban areas	^^	^
Migration	^^	^
Intensification of land use	^^	^
Biodiversity	vv	v
Threat of submergence	^^	N/A
Household adaptation	?	?
Engineering interventions	^^	^

^ increasing; ^^ strongly increasing; v decreasing; vv strongly decreasing; ? no data/unclear; N/A—not applicable

Fig. 1.1 Some delta features in the Anthropocene. **a** Natural environments—The Sunderbans, a world heritage site, India. **b** A delta city, Khulna, Bangladesh. **c** Intensive aquaculture, Bangladesh. **d** Cyclone evacuation centre, Bangladesh. **e** Agriculture—Polders, Bangladesh. **f** Major infrastructure—Akosombo Dam, upstream of the Volta Delta, Ghana. **g** Adaptation—A major storm surge barrier—the Maeslantkering, The Netherlands. **h** A world delta city, Shanghai, China (Photos: **a**, **b**, **e** Robert J Nicholls; **c**, **d** Attila N Lázár; **f**, **g**, **h** reprinted under licence CC BY-SA 1.0 and 3.0)

more than 500 million people live in and around deltas globally: or seven per cent of the human population on one per cent of global land area (Ericson et al. 2006; Woodroffe et al. 2006). Most large delta populations are found in mid and low latitude deltas in the global South associated with large rural populations depending on agriculture for their livelihoods, especially in Asia. Many deltas are also associated with large and rapidly growing mega-cities such as Cairo, Dhaka, Kolkata and Shanghai.

With their large populations and economic opportunities, deltas are a key focus for development. Deltas have also been associated with the threat of climate-induced sea-level rise and subsidence and there is concern this could act as a brake on economic development (de Souza et al. 2015; Hallegatte et al. 2016). Due to the low elevation of deltas, small changes in water level can have profound hydrological effects, including inundation, salinity and waterlogging with severe impacts on livelihoods. In the extreme, the spectre of mass forced migration from deltas due to sea-level rise was raised 30 years ago (Milliman et al. 1989). However, while such projections have become received wisdom and widely repeated (Gemenne 2011), there has been little systematic scientific investigation of demographic realities of future population movements and settlement patterns. Most integrated analyses show the complexity of delta processes and the occurrence of multiple and interacting drivers of change (Tessler et al. 2015; Nicholls et al. 2016). This book examines the current and future trajectories of Anthropocene deltas by focussing on the full range of interrelated environmental and social dynamics. It aims to understand the opportunities as well as the threats in deltas under Anthropocene conditions.

1.2 Trends in Deltas, Their Catchments and Adjacent Areas

There are deltas in every inhabited continent in the world, as illustrated in Fig. 1.2. There is a strong concentration of densely populated deltas in south, south-east and east Asia. Based on Woodroffe et al. (2006), the largest delta by population is the

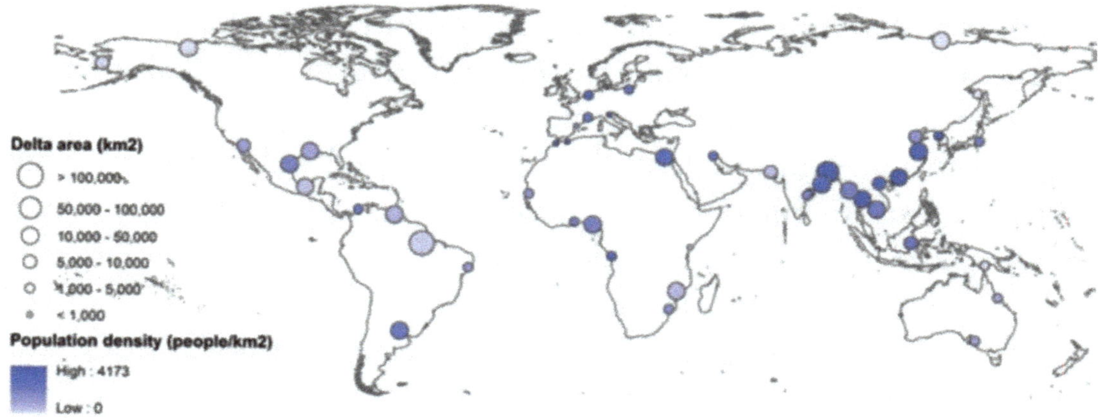

Fig. 1.2 A map of 47 global deltas by size and population density (based on data in Dunn 2017)

Ganges-Brahmaputra-Meghna (GBM) Delta (India and Bangladesh) with more than 130 million inhabitants. The other eight significant and populous deltas around the Himalayas, the Irrawaddy (Myanmar), Chao Phayra (Thailand), Mekong (Vietnam), Song Hong (Red) (Vietnam), Pearl (China), Changjiang (China), Huanghe (China) and the abandoned Huanghe (Jiangsu) (China); an additional 195 million people, living at densities of 400–5000 people/km^2. Further large delta populations are found along the Indian coast of the Bay of Bengal coast, for example the Mahanadi, as well as in Indonesia where many small deltas exist and coalesce into a coastal plain with dense populations, as in Java. Outside Asia there are deltas in Europe, the Netherlands is a delta country; North America, the Mississippi is an iconic delta; and South America, where the world's largest delta in area, the Amazon, is located. Africa has the vital Nile Delta, which supports around 40 million people including the city of Cairo, and a number of smaller locally important deltas such as the Volta in Ghana.

Deltas have been changing rapidly over the last few decades. They naturally evolve by the interplay of sediment supply, redistribution and loss, including the natural subsidence that occurs in all deltas. Hence, some deltas may experience accretion in certain areas and erosion elsewhere related to the local sediment budget (see Chapter 2). In the Anthropocene, a range of additional processes induced by humans have operated across scales as summarised in Table 1.2.

Table 1.2 Major human processes shaping delta evolution by scale

Scale	Human-induced process	Sources/information in this book
Global	Climate change and sea-level rise	Chapter 6
	Globalisation, trade and world markets	Chapter 8
Catchment	Land use change and sediment yield	Syvitski et al. (2009); Chapter 5
	Dams construction and increased water (and sediment) retention	
Delta	Enhanced subsidence (due to ground fluid with-drawal and/or drainage)	Ericson et al. (2006), Syvitski et al. (2009), Tessler et al. (2018); Chapter 5
	Enhanced flood protection (stopping sedimentation)	Syvitski et al. (2009); Chapters 5, 9 and 10
	Urbanisation and associated infrastructure	Szabo et al. (2016a); Chapters 7–9
	Intensified agriculture and aquaculture	Chapters 6, 8 and 9

At the global scale, human-induced climate change threatens deltas with accelerating sea-level rise, with more intense storms being of particular concern. These effects will be much larger in the future. Indirectly, deltas are also exposed to global markets and trends and this can drive changes at smaller scales, such as the growth in shrimp aquaculture for export. At the regional and catchment scale, human changes in the catchments, such as dams, can change the delivery of water and reduce the supply of sediment, sometimes almost completely. Hydrological effects in deltas are also apparent with issues such as intensified saltwater intrusion during the dry season reflecting multiple factors including upstream changes, delta changes and sea-level rise (e.g. Nicholls et al. 2018b). Deltas have long been important food producers reflecting their significant productive ecosystem services. With growing populations, agriculture has been greatly intensified in deltas for local and wider use, as has aquaculture. This is widely associated with polderisation where low-lying areas are surrounded by dykes and drainage is improved, allowing more easy regulation of water levels and excluding natural flooding and sedimentation. Agriculture intensification leads to less subsistence agriculture and lower agriculture employment (e.g. Amoako-Johnson et al. 2016). Agricultural systems often require intensive irrigation infrastructure to manage water during dry periods. Drainage and fluid withdrawal leads to accelerated subsidence and falling land levels increase the likelihood of waterlogging and flooding, driving the construction or upgrade of flood defences. Most populous deltas in the most recent decades have also experienced major rural to urban migration and growing urban centres, within or adjacent to the delta (Szabo et al. 2016a). The urbanisation trend has led to major infrastructure demands including enhanced urban flood management and protection. Consequently, deltas are widely moving towards being highly engineered landscapes and are now a long way from their previous natural or semi-natural state.

Figure 1.3 contrasts an archetypal pristine versus an Anthropocene delta, emphasising the intensification of a wide range of human activities in the Anthropocene delta. In the real world, deltas are at different stages of these processes, but all deltas with significant populations, it is argued, are following similar trajectories. The incremental nature of many of the trends is also apparent: they accumulate with time such as sea-level rise, loss of elevation and growth of flood defences (Tessler et al. 2015). There is also a strong coupling between natural processes and engineering responses that can lead to co-evolution and lock-in (Welch et al. 2017). Hence, deltas are complex systems and historical precedent is no longer a guide to the future. Adaptation and flexibility to the seasons and to climate extremes have always been necessary features of successful human occupancy of deltas. However, many historic adaptation behaviours may no longer be fit for purpose, due to changing pressures and climatic changes outside of lived experience.

For the future, climate change and sea-level rise have long been recognised as a major threat to deltas (Milliman et al. 1989; Tsyban et al. 1990). However, deltas in the Anthropocene are influenced by much more than climate change. Catchment changes, large-scale engineering interventions in rivers and coasts, population dynamics and socio-ecological interactions dominate virtually all observed changes in the major deltas of the world to date. These factors can only be compounded by climate change and sea-level rise which is a growing issue, especially after 2050 when

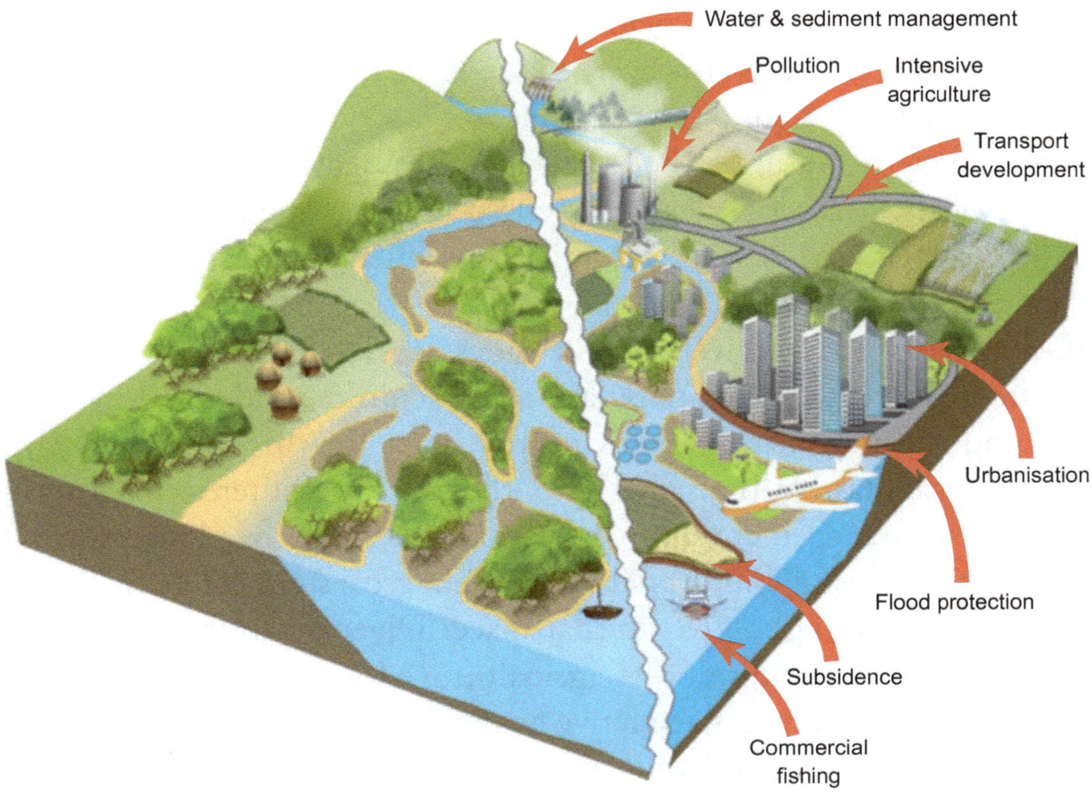

Fig. 1.3 A pristine (left) and an Anthropocene (right) archetypal delta, emphasising the intensification of a wide range of human activities within deltas during the Anthropocene

changes could be dramatic (Brown et al. 2018; Nicholls et al. 2018a). Consequently, as the Anthropocene continues, societies will be transformed and will need to address new and emerging threats and challenges. In many ways, delta societies and regions epitomise the challenges found elsewhere: the lessons drawn from deltas can inform the needs in other settings.

1.3 Possible Consequences of Changing Deltas

Current and future Anthropocene trends and trajectories have important consequences for the deltas and their populations. For example, the geomorphic trend of loss of elevation leads to potential submergence of the deltas and land loss or a growing dependence on flood defences and pumped drainage (Blum and Roberts 2009; Syvitski et al. 2009). This trend is apparent in many deltas around the world, related to subsidence and sea-level rise. However, as sea-level rise accelerates and compensating sediment supplies continue to fall, an existential threat to deltas is likely to occur in the Anthropocene. Relative sea-level rise also has important hydrological trends, especially waterlogging and salinisation (Payo et al. 2017) with potential consequences for agriculture even if there are flood defences.

Deltas have rich biodiversity and wider ecosystem service values. The pressure from expanding population, agriculture, industrial development and submergence is removing the space for this important characteristic of deltas. At the same time, there is more interest in 'working with nature' management approaches, such as mangrove buffers, which might give space for biodiversity and their ecosystem services. But are delta societies willing to give up that space? Diversification to high value crops has increased the value of agriculture. Moreover, subsistence agriculture is increasingly less attractive as the economies diversify and relative wages and returns to labour in primary industries stagnate. Climate change is also making traditional agriculture more risky and less attractive, suggesting a trend to decline in employment in agriculture and consolidation of agricultural holdings. These trends in the nature of agriculture and land use represent a profound change to deltas globally. Moreover, dyke-based delta management and the move away from annual floods may undermine agricultural economics due to the high cost of fertiliser, as shown in the Mekong Delta (Chapman and Darby 2016). It is not clear who will benefit from these agriculture transitions or from rapid diversification of rural economies in delta regions and urbanisation. For example, the economy of Bangladesh is currently growing at more than seven per cent, and over the past 20 years the economy has increased five times, according to World Bank data. These observed changes indicate the nature of the delta economies are likely to continue to change fundamentally in the coming decades.

In addition, there are major demographic transitions in societies in most countries with large populous deltas (Szabo et al. 2016a). This suggests the emergence of ageing and stable and even falling populations in the future. The growth of cities in and around the deltas has been profound and reflects a large migration of people from rural areas.

Data from observations and projections of populations of selected large cities from 1950 to 2020 (Table 1.3) shows the largest growth in the GBM Delta, especially in Dhaka, Kolkata and Chattogram (also known as Chittagong), and in relative terms Dhaka, Bhubaneswar and Lomé. Continued substantial growth is projected from 2020 to 2035 in all the cities, except Khulna. When considering the effects of climate change on migration, this demonstrates existing substantial high levels of mobility and population dynamics in many deltas.

1.4 Policy Implications

The sustainability of deltas in the Anthropocene requires the long term health of physical and ecological systems alongside widely distributed development that leaves no-one behind. Day et al. (2016) consider delta sustainability within the context of global biophysical and socio-economic constraints. They state that geomorphic, ecological and economic aspects of the delta sustainability are strongly influenced by society. However, critical factors of a sustainable delta society include livelihood sustainability, demography, well-being and critically their trajectories through time. The process of managing deltas will require an adaptive approach to the development of policy with an effective strategy for monitoring progress, scenario development and learning. Taken together

Table 1.3 Selected observed and projected city populations (thousands) in and adjacent to deltas from 1950 to 2035 (Data from UN DESA [2018]) (see Fig. 1.4 and Table 1.4)

City	1950	1970	1990	2010	2020	2035
Accra	177	631	1197	2060	2514	3632
Lomé	33	192	619	1466	1828	2947
Bhubaneswar	16	98	395	868	1163	1649
Kolkata	4604	7329	10,974	14,003	14,850	19,564
Khulna	61	310	985	1098	954	1213
Dhaka	336	1374	6621	14,731	21,006	31,234
Chattogram (Chittagong)	289	723	2023	4106	5020	7110

these diverse elements represent the challenges of delta planning in the Anthropocene.

The complexity and intensity of the changes discussed in Sects. 1.2 and 1.3 suggests that management and development in deltas will be challenging. Inevitably important choices and trade-offs will emerge and there will be a need for clear priorities and goals. This is apparent when the aspirations of the UN Sustainable Development Goals (SDGs) are considered, as the issues of trade-offs and choice and how these challenges are addressed have received little attention (Szabo et al. 2016b; Hutton et al. 2018). While delta residents have had to be highly adaptive to exploit delta livelihoods, all the changes above suggest that traditional adaptation will be insufficient and important new adaptation efforts linked to delta development will be required in the future.

The scale of the challenge for delta management in the Anthropocene would suggest that new and more integrated approaches are required. Deltas are inherently challenging systems to manage due to the high sensitivity to multiple drivers and the potential for co-evolution and lock-in (Welch et al. 2017; Seijger et al. 2018). As an example, construction of flood defences and drainage attracts development, excludes sedimentation and promotes subsidence of organic soils, increasing the consequences of flooding if the defences fail. Hence, construction of flood defences ultimately promotes bigger and higher defences and the consequences of failure grow. Hallegatte et al. (2013) analysed flooding in coastal cities and found protection resulted in fewer, but bigger, disasters in terms of economic damage. This effect was largest in cities located in deltas as they subside, in addition to experiencing climate-induced sea-level rise. More innovative ways of managing deltas are needed such as allowing controlled flooding and sedimentation. This approach has been used in the Yellow River, China (Han et al. 1995) and is applied at a small scale as so-called Tidal River Management in Bangladesh (Chapters 2, 6 and 10). A major dilemma for such strategies is whether they can be scaled up for Anthropocene conditions when sediment supply from feeding rivers is widely diminishing (e.g. Rogers and Overeem 2017; Dunn et al. 2018).

Whole delta plans aim to consider the entire delta and adopt integrated perspectives. They are proliferating. In the Mississippi Delta, coastal restoration is actively focussing on ecosystem restoration and coastal safety (CPRAL 2017). In the Netherlands, the Dutch Delta Commission has been established and similar plans are being developed in the Mekong and Bangladesh (Seijger et al. 2017). In all these examples, there is a vision of coordinated delta development and adaptive approaches to an uncertain future.

1.5 The Book Approach and Structure

This book assesses the influences and interactions in deltas in order to appraise the sustainability and potential futures for deltas in the Anthropocene. It draws significantly from research conducted in the multi-disciplinary Deltas, Vulnerability and Climate Change: Migration and Adaptation project that researched a diversity of the world's deltas from 2014 to 2018 (DECCMA 2018). The DECCMA initiative generated scenarios of downscaled climate change and sea-level rise; analysed hotspots of vulnerability and risk; examined individual behaviour around adaptation practices and migration patterns and consequences; developed land cover and national agro-ecological zone production; and modelled the macroeconomy of deltas with projections of how environmental change would affect both the scale of economic activity and demand for labour. The book, building on findings of the DECCMA consortium, seeks to answer a number of questions. What are the key characteristics of the Anthropocene economic transition in deltas? What are the implications of this transition for twenty-first-century management and adaptation of deltas, including the wider policy implications? Do deltas offer an early insight into the impact of the Anthropocene transition of relevance to the management for other socio-ecological contexts?

The chapters and individual assessments draw on new empirical insights from DECCMA based on primary data and analysis for three significant and contrasting deltas: the world's largest and most populous delta, the GBM, spanning India and Bangladesh, the Mahanadi, entirely within India and the Volta in West Africa (Fig. 1.4). They also draw on a renewed and vigorous global community of researchers

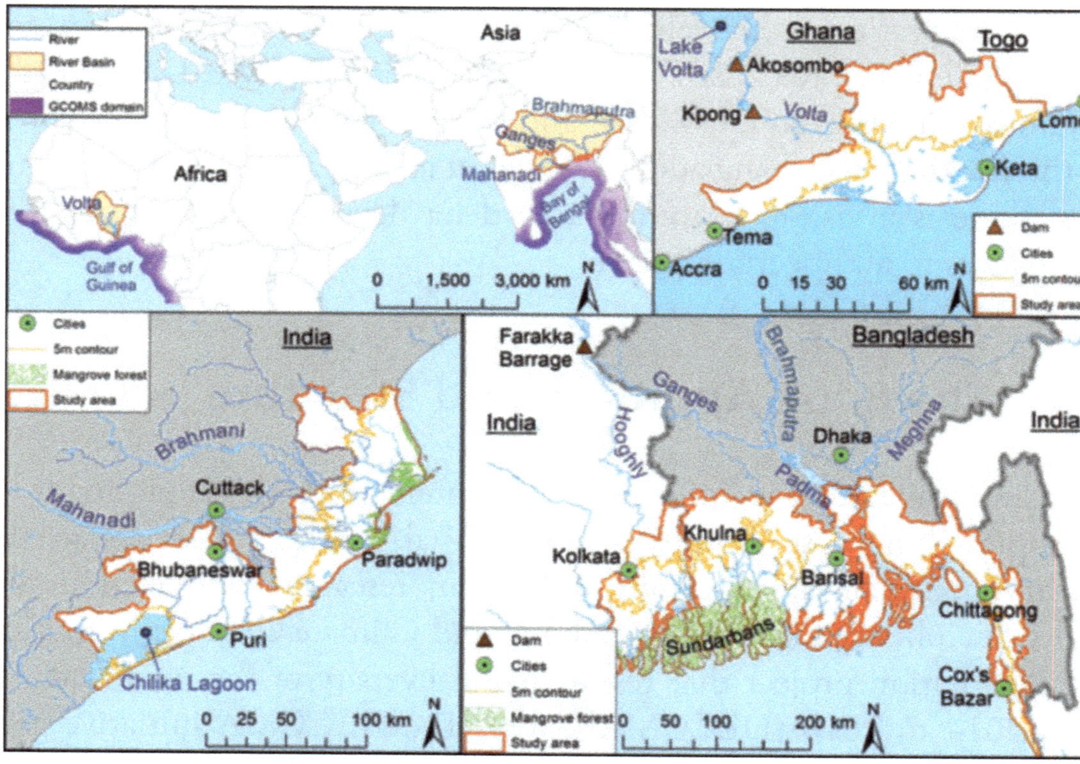

Fig. 1.4 Delta regions considered, including their associated catchments and the shelf sea regions (analysed for fisheries): **a** Volta Delta. **b** Mahanadi Delta. **c** Ganges-Brahmaputra-Meghna Delta (Modified from Kebede et al. [2018] under CC BY 4.0)

seeking to support delta planning through integrated science. Deltas can be defined in various ways, such as the location of the most upstream distributary, or the surface area composed of Holocene sediments (e.g. Woodroffe et al. 2006). The principal focus of analysis in this book is on the coastal portion of deltas. Coastal parts of deltas are the crucible of the impacts of climate change, especially sea-level rise and subsidence, as well as areas with high population density and economic activity. The study areas for the DECCMA deltas were defined as a subset of the physical delta limited to land below the five-metre contour. The specific study area was defined by selecting all the administrative units within this physical delta, i.e. the political units where these impacts and adaptation are experienced and represent dilemmas and trade-offs. In Bangladesh, a slightly different approach was taken and the coastal zone, as recognised by the Government of Bangladesh, was analysed. This excludes large areas in Bangladesh below the five-metre contour further inland, but gives an administratively meaningful study area. This is also the area subject to tropical cyclones. Some of the key characteristics of the delta study areas are given in Table 1.4.

The population density is lowest in the Volta and exceeds 1000 people/km^2 in the dominantly rural setting of the GBM Delta. The deltas all contain large and growing cities, or are adjacent to such cities, with Accra and Lomé on either side of the Volta Delta. Agriculture and fisheries are important components of the economy, especially employment. However, the economies are more diverse than often considered, with large industry and construction, and especially service components, although these aspects of the economy have a strong linkage to agriculture and fisheries.

The multi-disciplinary, policy-orientated research reported throughout the book explores the effects of a range of environmental and economic scenarios on migration, adaptation, governance, as well as asset poverty, agricultural and fisheries productivity and nutritional levels. The research is strongly aligned to the United Nation's SDGs as delta populations often experience extremes of poverty, gender and structural inequality, variable levels of health and well-being, while being vulnerable to extreme and systematic environmental and climate change.

The book is structured as follows. First, the three deltas are considered in detail in Chapters 2–4: the GBM, Mahanadi and Volta Deltas, respectively. They ask the questions, how are these deltas today, and what does the Anthropocene mean for each of them? Six thematic chapters then explore challenges of development and delta futures, based on global evidence and drawing from the DECCMA example deltas in particular. Chapter 5 considers fluvial sediment supply and relative sea-level change in deltas. This is the process which produced and sustained the world's deltas over the last few millennia and centuries. However, in the Anthropocene the sediment supply is failing due to upstream change, especially the construction of dams. This suggests intensified hazards and growing land loss, or a growing dependence on dykes and polders. Only radical management in the deltas and in the catchments can change this trend. Chapter 6 examines the hazards, exposure, vulnerability and risks within deltas, and their spatial expression, including recognising hotspots. These concepts have emerged

Table 1.4 Economic and demographic characteristics of the selected delta regions

Delta region	Total area (km^2) and area below 5 m elevation (%)	Population and population density (people/km^2)	Large cities	GDP (and employment) by sector			GDP per capita (2012, USD, power purchasing parities)
				Agriculture and fishing	Industry and construction	Services	
Volta	5136 (38)	0.86×10^6 (168)	None in delta region. Nearby Accra and Lomé, Togo	29% (40%)	31% (22%)	40% (38%)	1048
Mahanadi	12,856 (38)	8.1×10^6 (625)	Bhubaneswar	29% (44%)	17% (17%)	54% (39%)	1958
Ganges-Brahmaputra-Meghna	51,493 (60)	56.1×10^6 (1101)	Kolkata, Khulna and Chattogram, with Dhaka nearby	15% (26%)	38% (11%)	47% (63%)	1847

over the last few decades as key information for environmental policy analysis. Hotspots also inform other analysis and helped to guide the analysis of migration in Chapter 7 and adaptation in Chapter 9. Chapter 7 assesses where people live and move within deltas using analysis of the census and dedicated household surveys. It finds that there is high human mobility in deltas today driven largely by economic reasons, including movement to cities. Hence, any migration due to environmental change will be in addition to the existing migration processes. Chapter 8 focuses on delta economics and sustainability. While agriculture is a large part of the economy, and provides much of the employment, the three delta economies have growing service, trade-transport, industry and construction sectors. Fisheries are also important. Nonetheless, the delta economies are quite vulnerable to climate change impacts in the next few decades. Chapter 9 considers adaptation to change as a response to these and other threats. Adaptation is already widespread in deltas. Much of the adaptation activity at household level focusses on reducing past and present vulnerabilities, with little attention to longer term risks. Government adaptation initiatives often address structural needs and managing the dynamic nature of these environments. There is no joined up vision of adaptation in deltas as yet, although future adaptation offers delta societies opportunities and choices to deal with the challenges of the Anthropocene, and it is important to recognise that adaptation choices will shape the future evolution of deltas. Chapter 10 assesses adaptation at the delta scale, and the role of trade-offs and plausible development pathways. Delta management in the Anthropocene will involve the consideration of trade-offs and the balancing of positive and negative outcomes for delta functions and the societies that rely on them. Using integrated models, the chapter demonstrates that trade-offs are crucial governance challenges for the future sustainability of deltas, probably illustrating a wider challenge for the Anthropocene. Lastly, Chapter 11 synthesises the book, considering the questions posed for deltas in the Anthropocene. The trends emerging in each chapter are reviewed and considered together, and possible trajectories for the Volta, Mahanadi and GBM deltas are presented. Using these insights, it considers the notion of a sustainable delta from a variety of perspectives, and what this might mean under Anthropocene conditions.

These chapters examine deltas from a range of biophysical and socio-economic perspectives. The consideration of migration and population in deltas in Chapters 7 and 9, for example, uses a new and bespoke survey on migration and adaptation patterns across the DECCMA deltas: the survey involved more than 7500 households and individual migrants in rural parts of the deltas and in their destination cities (DECCMA 2018). Such data on migration and household-level adaptation to hazards and climate changes provides important new insights on delta demography and human dilemmas in Anthropocene deltas. Analysis of the delta economy in Chapter 8 uses newly developed macroeconomic models for the DECCMA deltas that allows direct comparative analysis of the macro-level trends and impacts of environmental change on economic activity. Synthesis across all these dimensions provides important new insights on key challenges and trade-offs for delta societies in the Anthropocene.

References

Amoako-Johnson, F., Hutton, C. W., Hornby, D., Lázár, A. N., & Mukhopadhyay, A. (2016). Is shrimp farming a successful adaptation to salinity intrusion? A geospatial associative analysis of poverty in the populous Ganges–Brahmaputra–Meghna Delta of Bangladesh. *Sustainability Science, 11*(3), 423–439. https://doi.org/10.1007/s11625-016-0356-6.

Bianchi, T. S. (2016). *Deltas and humans: A long relationship now threatened by global change*. Oxford, UK: Oxford University Press.

Blum, M. D., & Roberts, H. H. (2009). Drowning of the Mississippi Delta due to insufficient sediment supply and global sea-level rise. *Nature Geoscience, 2*, 488. https://doi.org/10.1038/ngeo553.

Brown, S., Nicholls, R. J., Lázár, A. N., Hornby, D. D., Hill, C., Hazra, S., et al. (2018). What are the implications of sea-level rise for a 1.5, 2 and 3 °C rise in global mean temperatures in the Ganges-Brahmaputra-Meghna and other vulnerable deltas? *Regional Environmental Change, 18*(6), 1829–1842. https://doi.org/10.1007/s10113-018-1311-0.

Chapman, A., & Darby, S. (2016). Evaluating sustainable adaptation strategies for vulnerable mega-deltas using system dynamics modelling: Rice agriculture in the Mekong Delta's An Giang Province. *Vietnam. Science of the Total Environment, 559,* 326–338. https://doi.org/10.1016/j.scitotenv.2016.02.162.

CPRAL. (2017). *Louisiana's comprehensive master plan for a sustainable coast: Commited to our coast.* Baton Rouge, LA: Coastal Protection and Restoration Authority of Louisiana. http://coastal.la.gov/wp-content/uploads/2017/04/2017-Coastal-Master-Plan_Web-Single-Page_CFinal-with-Effective-Date-06092017.pdf. Last accessed 20 December 2018.

Day, J. W., Agboola, J., Chen, Z., D'Elia, C., Forbes, D. L., Giosan, L., et al. (2016). Approaches to defining deltaic sustainability in the 21st century. *Estuarine, Coastal and Shelf Science, 183,* 275–291. https://doi.org/10.1016/j.ecss.2016.06.018.

DECCMA. (2018). *Climate change, migration and adaptation in deltas: Key findings from the DECCMA project* (Deltas, Vulnerability and Climate Change: Migration and Adaptation [DECCMA] Report). Southampton, UK: DECCMA Consortium. https://www.preventionweb.net/publications/view/61576. Last accessed 27 November 2018.

de Souza, K., Kituyi, E., Harvey, B., Leone, M., Murali, K. S., & Ford, J. D. (2015). Vulnerability to climate change in three hot spots in Africa and Asia: Key issues for policy-relevant adaptation and resilience-building research. *Regional Environmental Change, 15*(5), 747–753. https://doi.org/10.1007/s10113-015-0755-8.

Donges, J. F., Winkelmann, R., Lucht, W., Cornell, S. E., Dyke, J. G., Rockström, J., et al. (2017). Closing the loop: Reconnecting human dynamics to Earth System science. *The Anthropocene Review, 4*(2), 151–157. https://doi.org/10.1177/2053019617725537.

Dunn, F. E. (2017). *Multidecadal fluvial sediment fluxes to major deltas under environmental change scenarios: Projections and their implications* (PhD thesis). Faculty of Geography and the Environment, University of Southampton, Southampton, UK.

Dunn, F. E., Nicholls, R. J., Darby, S. E., Cohen, S., Zarfl, C., & Fekete, B. M. (2018). Projections of historical and 21st century fluvial sediment delivery to the Ganges-Brahmaputra-Meghna, Mahanadi, and Volta deltas. *Science of the Total Environment, 642,* 105–116. https://doi.org/10.1016/j.scitotenv.2018.06.006.

Ericson, J. P., Vörösmarty, C. J., Dingman, S. L., Ward, L. G., & Meybeck, M. (2006). Effective sea-level rise and deltas: Causes of change and human dimension implications. *Global and Planetary Change, 50*(1–2), 63–82. https://doi.org/10.1016/j.gloplacha.2005.07.004.

Gemenne, F. (2011). Why the numbers don't add up: A review of estimates and predictions of people displaced by environmental changes. *Global Environmental Change, 21,* S41–S49. https://doi.org/10.1016/j.gloenvcha.2011.09.005.

Goudie, A. S., & Viles, H. A. (2016). *Geomorphology in the Anthropocene.* Cambridge, UK: Cambridge University Press.

Hallegatte, S., Bangalore, M., Bonzanigo, L., Fay, M., Kane, T., Narloch, U., et al. (2016). *Shock waves: Managing the impacts of climate change on poverty.* Washington, DC: World Bank. https://openknowledge.worldbank.org/handle/10986/22787. License: CC BY 3.0 IGO.

Hallegatte, S., Green, C., Nicholls, R. J., & Corfee-Morlot, J. (2013). Future flood losses in major coastal cities. *Nature Climate Change, 3*, 802–806. https://doi.org/10.1038/nclimate1979.

Han, M., Wu, L., Hou, C., & Liu, G. (1995). Sea-level rise and the North China coastal plain: A preliminary analysis. *Journal of Coastal Research, S14,* 132–150.

Hutton, C. W., Nicholls, R. J., Lázár, A. N., Chapman, A., Schaafsma, M., & Salehin, M. (2018). Potential trade-offs between the Sustainable Development Goals in Coastal Bangladesh. *Sustainability, 10*(4), 1008. http://dx.doi.org/10.3390/su10041108.

Ibáñez, C., Alcaraz, C., Caiola, N., Prado, P., Trobajo, R., Benito, X., et al. (2019). Basin-scale land use impacts on world deltas: Human vs natural forcings. *Global and Planetary Change, 173,* 24–32. https://doi.org/10.1016/j.gloplacha.2018.12.003.

Kebede, A. S., Nicholls, R. J., Allan, A., Arto, I., Cazcarro, I., Fernandes, J. A., et al. (2018). Applying the global RCP–SSP–SPA scenario framework at sub-national scale: A multi-scale and participatory scenario approach. *Science of the Total Environment, 635,* 659–672. https://doi.org/10.1016/j.scitotenv.2018.03.368.

Milliman, J. D., Broadus, J. M., & Gable, F. (1989). Environmental and economic implications of rising sea level and subsiding deltas: The Nile and Bengal examples. *Ambio, 18*(6), 340–345.

Nicholls, R. J., Brown, S., Goodwin, P., Wahl, T., Lowe, J., Solan, M., et al. (2018a). Stabilization of global temperature at 1.5°C and 2.0°C: Implications for coastal areas. *Philosophical Transactions of the Royal Society, 376*(2119). https://doi.org/10.1098/rsta.2016.0448.

Nicholls, R. J., Hutton, C. W., Adger, W. N., Hanson, S. E., Rahman, M. M., & Salehin, M. (Eds.). (2018b). *Ecosystem services for well-being in deltas: Integrated assessment for policy analysis*. London, UK: Palgrave Macmillan.

Nicholls, R. J., Hutton, C. W., Lázár, A. N., Allan, A., Adger, W. N., Adams, H., et al. (2016). Integrated assessment of social and environmental sustainability dynamics in the Ganges-Brahmaputra-Meghna Delta, Bangladesh. *Estuarine and Coastal Shelf Science, 183,* 370–381. https://doi.org/10.1016/j.ecss.2016.08.017.

Payo, A., Lázár, A. N., Clarke, D., Nicholls, R. J., Bricheno, L., Mashfiqus, S., et al. (2017). Modeling daily soil salinity dynamics in response to agricultural and environmental changes in coastal Bangladesh. *Earth's Future, 5*(5), 495–514. https://doi.org/10.1002/2016EF000530.

Renaud, F. G., Syvitski, J. P. M., Sebesvari, Z., Werners, S. E., Kremer, H., Kuenzer, C., et al. (2013). Tipping from the Holocene to the Anthropocene: How threatened are major world deltas? *Current Opinion in Environmental Sustainability, 5*(6), 644–654. https://doi.org/10.1016/j.cosust.2013.11.007.

Ribot, J. (2014). Cause and response: Vulnerability and climate in the Anthropocene. *Journal of Peasant Studies, 41*(5), 667–705. https://doi.org/10.1080/03066150.2014.894911.

Rogers, K. G., & Overeem, I. (2017). Doomed to drown? Sediment dynamics in the human-controlled floodplains of the active Bengal Delta. *Elementa Science of the Anthropocene, 5*, 65. https://doi.org/10.1525/elementa.250.

Seijger, C., Douven, W., van Halsema, G., Hermans, L., Evers, J., Phi, H. L., et al. (2017). An analytical framework for strategic delta planning: Negotiating consent for long-term sustainable delta development. *Journal of Environmental Planning and Management, 60*(8), 1485–1509. https://doi.org/10.1080/09640568.2016.1231667.

Seijger, C., Ellen, G. J., Janssen, S., Verheijen, E., & Erkens, G. (2018). Sinking deltas: Trapped in a dual lock-in of technology and institutions. *Prometheus, 35*, 1–21. https://doi.org/10.1080/08109028.2018.1504867.

Steffen, W., Persson, Å., Deutsch, L., Zalasiewicz, J., Williams, M., Richardson, K., et al. (2011). The Anthropocene: From global change to planetary stewardship. *AMBIO: A Journal of the Human Environment, 40*(7), 739–761. https://doi.org/10.1007/s13280-011-0185-x.

Syvitski, J. P. M. (2008). Deltas at risk. *Sustainability Science, 3*(1), 23–32. https://doi.org/10.1007/s11625-008-0043-3.

Syvitski, J. P. M., Kettner, A. J., Overeem, I., Hutton, E. W. H., Hannon, M. T., Brakenridge, G. R., et al. (2009). Sinking deltas due to human activities. *Nature Geoscience, 2*(10), 681–686. https://doi.org/10.1038/ngeo629.

Szabo, S., Adger, W. N., & Matthews, Z. (2018). Home is where the money goes: Migration-related urban-rural integration in delta regions. *Migration and Development, 7*(2), 163–179. https://doi.org/10.1080/21632324.2017.1374506.

Szabo, S., Brondizio, E., Renaud, F. G., Hetrick, S., Nicholls, R. J., Matthews, Z., et al. (2016a). Population dynamics, delta vulnerability and environmental change: Comparison of the Mekong, Ganges-Brahmaputra and Amazon delta regions. *Sustainability Science, 11*(4), 539–554. https://doi.org/10.1007/s11625-016-0372-6.

Szabo, S., Nicholls, R. J., Neumann, B., Renaud, F. G., Matthews, Z., Sebesvari, Z., et al. (2016b). Making SDGs work for climate change hotspots. *Environment: Science and Policy for Sustainable Development, 58*(6), 24–33. https://doi.org/10.1080/00139157.2016.1209016.

Tessler, Z. D., Vörösmarty, C. J., Grossberg, M., Gladkova, I., Aizenman, H., Syvitski, J., et al. (2015). Profiling risk and sustainability in coastal deltas of the world. *Science, 349*(6248), 638–643. https://doi.org/10.1126/science.aab3574.

Tessler, Z. D., Vörösmarty, C. J., Overeem, I., & Syvitski, J. P. M. (2018). A model of water and sediment balance as determinants of relative sea level rise in contemporary and future deltas. *Geomorphology, 305,* 209–220. https://doi.org/10.1016/j.geomorph.2017.09.040.

Tsyban, A. V., Everett, J. T., & Titus, J. G. (1990). World oceans and coastal zones. In W. Tegart, G. W. Sheldon, & C. Griffiths (Eds.), *Climate change: The IPCC impacts assessment*. Canberra, Australia: Australian Government Publishing Service. http://papers.risingsea.net/federal_reports/IPCC-far_wg_II_chapter_6.pdf. Last accessed 21 January 2019.

UN DESA. (2018). *World Urbanization Prospects: The 2018 Revision* (Online ed.). Department of Economic and Social Affairs, Population Division, United Nations. https://population.un.org/wup/Download/. Last accessed 2 January 2019.

Verburg, P. H., Dearing, J. A., Dyke, J. G., Leeuw, S., Seitzinger, S., Steffen, W., et al. (2016). Methods and approaches to modelling the Anthropocene. *Global Environmental Change, 39,* 328–340. https://doi.org/10.1016/j.gloenvcha.2015.08.007.

Welch, A. C., Nicholls, R. J., & Lázár, A. N. (2017), Evolving deltas: Co-evolution with engineered interventions. *Elementa Science of the Anthropocene, 5*, 49. https://doi.org/10.1525/elementa.128.

Woodroffe, C. N., Nicholls, R. J., Saito, Y., Chen, Z., & Goodbred, S. L. (2006). Landscape variability and the response of Asian megadeltas to environmental change. In N. Harvey (Ed.), *Global change and integrated coastal management: The Asia-Pacific region* (pp. 277–314). New York, NY: Springer.

2

Ganges-Brahmaputra-Meghna Delta, Bangladesh and India: A Transnational Mega-Delta

Md. Munsur Rahman, Tuhin Ghosh, Mashfiqus Salehin, Amit Ghosh, Anisul Haque, Mohammed Abed Hossain, Shouvik Das, Somnath Hazra, Nabiul Islam, Maminul Haque Sarker, Robert J. Nicholls and Craig W. Hutton

2.1 The Ganges-Brahmaputra-Meghna Delta

The Ganges-Brahmaputra-Meghna (GBM) Delta at the north of the Bay of Bengal is administrated by both India and Bangladesh. It is characterised by a number of livelihood opportunities resulting from high population density, as well as a number of biophysical and socio-economic challenges (flooding, erosion, cyclones, salinisation, water logging, etc.) which are increasing alongside the changing climate and anthropogenic developments.

M. M. Rahman (✉) · M. Salehin · A. Haque · M. A. Hossain
Institute of Water and Flood Management, Bangladesh University of Engineering and Technology, Dhaka, Bangladesh
e-mail: mmrahman@iwfm.buet.ac.bd

T. Ghosh · A. Ghosh · S. Das · S. Hazra
School of Oceanographic Studies, Jadavpur University, Kolkata, India

N. Islam
Bangladesh Institute of Development Studies, Dhaka, Bangladesh

R. J. Nicholls et al. (eds.), *Deltas in the Anthropocene*,
https://doi.org/10.1007/978-3-030-23517-8_2

The people of this region are mainly dependent on the agricultural sector, while people living in the coastal belt are dependent on traditional monsoon rice cultivation as well as livelihood activities such as riverine and marine fishing and activities related to mangroves such as honey collection. Freshwater flooding is a common occurrence in the delta during the monsoon; it generates benefits such as increased soil fertility, aquifer recharge, replenished ecosystem and increased agricultural production. The delta also supports a diversity of ecosystem services that attract and support a large local population. One key area is the Sundarbans, the world's largest mangrove forest, covering 10,000 km^2 which is shared between Bangladesh (60%) and India (40%) (Fig. 2.1). The unique biodiversity of this area supports a diversity of livelihood options for the people living on its periphery (Gopal and Chauhan 2006).

The coastal population is exposed to climate hazards, including fluvio-tidal floods, tropical cyclones accompanied by storm surges, river bank erosion, salinity intrusion due to seasonal low flow levels in rivers and upstream water diversion, high levels of salinity in groundwater and arsenic contamination of shallow aquifers. Climate change and land use impacts are expected to reinforce many of these stresses (Dastagir 2015). These environmental stresses are believed to be enhancing already substantial displacement and migration. However, while the country has seen many planned and autonomous adaptations to minimise forced migration and displacement, situations often arise when people have little choice but to move (Mortreux et al. 2018). Consequently, for effective planning, it is important for policymakers

M. H. Sarker
Center for Environmental and Geographic Information Services, Dhaka, Bangladesh

R. J. Nicholls
School of Engineering, University of Southampton, Southampton, UK

C. W. Hutton
GeoData Institute, Geography and Environmental Science, University of Southampton, Southampton, UK

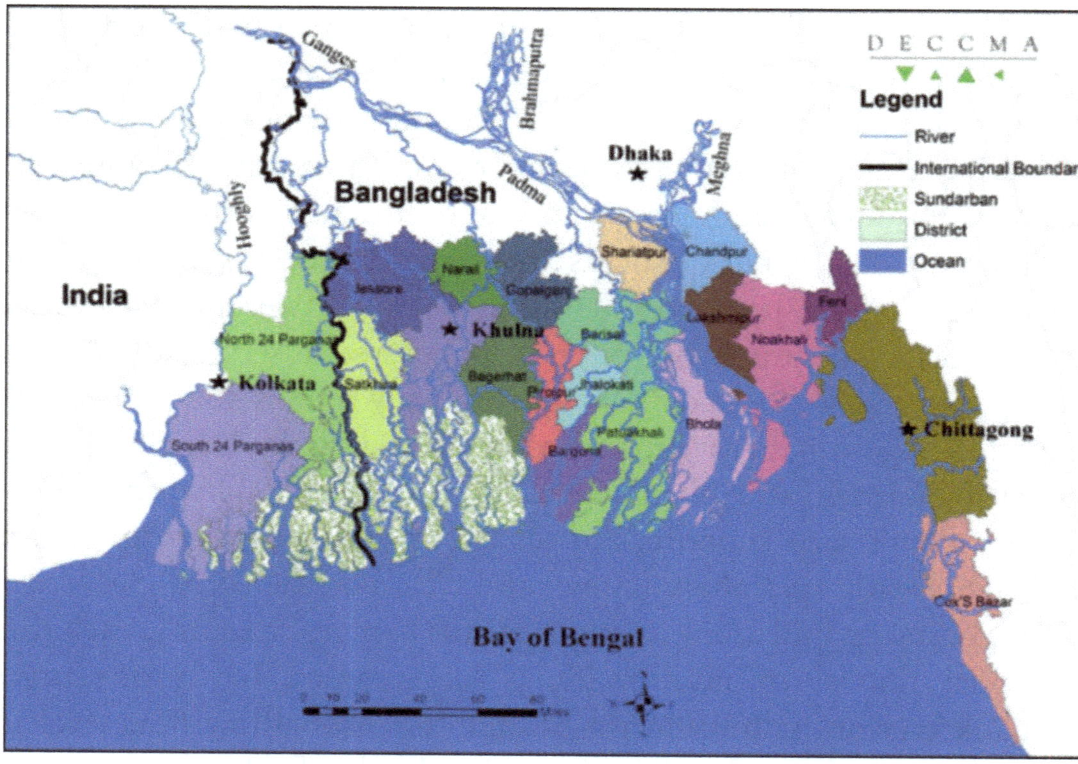

Fig. 2.1 The Ganges-Brahmaputra-Meghna Delta showing the coastal zone with administrative districts in both India and Bangladesh

to have an understanding of how effective adaptation options are, the circumstances under which people migrate, and if or when people see migration as an option in the context of available adaptation choices (Chapter 7).

Coastal Bangladesh has an extensive system of coastal embankments and polders built since the 1960s with the goal of reducing flooding/salinity, managing water levels and enhancing agriculture (Haque and Nicholls 2018). While the positive results from such interventions are visible in the form of increased agricultural production and enhanced regional connectivity, these benefits could not be translated into long-term gains across the GBM system (Noor 2018). Rather, these interventions are posing substantial challenges. Prolonged water logging due to silting up of river beds and hence reduced drainage capacity of floodplains, further exacerbated by ill-planned or ill-executed infrastructure projects, such as internal road system, water control infrastructure not being properly maintained and aquaculture and other economic activities obstructing drainage undermine the provision of sustainable services. In addition, they contributed to land subsidence, land use pattern changes and tidal influences on flooding (Islam et al. 2010).

The study area in this chapter considers the part of the delta that is most threatened by sea-level rise. In West Bengal, this corresponds to the political units that include areas below 5 m elevation near Kolkata—South 24 Parganas and North 24 Parganas (Fig. 2.1). In Bangladesh, the area below 5 m extends across much of the country and the study area corresponds to coastal Bangladesh as defined by the Government of Bangladesh. The study area includes part of the Kolkata metropolitan area, but excludes Dhaka: both these cities attract significant migration (see Chapter 1, Table 1.4). The population of the study area is 57 million and occupies more than 51,500 km^2. When compared to other papers such as Ericson et al. (2006) and Woodroffe et al. (2006) which consider the entire Holocene surface rising to 20 m elevation and a population exceeding 100 million people, it is apparent that the focus here is more coastal than in earlier analyses.

The aim of the chapter is to consider the whole of the coastal GBM Delta in India and Bangladesh from a biophysical and socio-economic perspective. There is a focus on the Anthropocene delta and its prospects and the consideration of both India and Bangladesh makes the chapter distinct compared to most analyses which are delimited by national boundaries. The chapter is structured as follows. The characteristics of delta-building and socio-ecological processes are outlined, followed by discussion on emerging opportunities and challenges, growth of settlements/land use, vulnerability mapping and options for adaptation, including migration.

2.2 Morphological Evolution of the Delta

The GBM Delta is a peripheral foreland basin formed through continent–continent collision formed over many millions of years (Raman et al. 1986). Physiographically, the GBM Delta can be divided into two major units—the Pleistocene uplands and the deltaic lowlands. There are four major terraces; two of these terraces flank the basin extending east of the Rajmahal hills and west of the Tripura Hills while the other two, Barind and Madhupur Forest, lie within the basin (Morgan and McIntire 1959). The other main physiographic division of the GBM Delta system is the Holocene alluvial plain and the delta.

Pleistocene eustatic sea-level fall has created widespread terraces and deep erosion of valleys by lowering of base level (Alam 1996). The sediments brought in by the Ganga-Brahmaputra system during the post-Pleistocene period appear to have mainly bypassed the delta and contributed to the rapid growth of Bengal deep-sea fan (Biswas 1992). Holocene sediments are found in the alluvial fans in the foothills of the Himalayas, the uplands such as the Tippera surface, the deep tectonic basin (Sylhet basin) and the GBM flood and delta plain, the most extensive unit of the GBM Delta.

Allison et al. (2003) examined the mineralogical properties of sediment for assessing sedimentary sequence resulting from the lower delta plain progradation in the late Holocene. From a series of 38 core sites across the delta, clay mineralogical and radiocarbon evidence agree that the lower delta plain progradation after the maximum transgression may have been in six phases as the Ganges and Brahmaputra grew together (Fig. 2.2). Clay mineralogy suggests an increasing influence of the Ganges in the upper section that may suggest a progradation of Ganges distributaries into the westernmost delta in the earliest phase (G1). The early (5000 cal years BP) deltas of the Brahmaputra (B1) and the Ganges were located far inland of the present shoreline, reflecting the large amount of accommodation space available in the tectonically active Bengal basin. Allison et al. (2003) also found that the shoreline progradation associated with the two rivers was separate from 5000 cal years BP until they merged into the present Meghna estuary as recently as about 200 years ago. A series of eastward steps of the Ganges occurred in three main phases (G1–G3). In each phase, delta progradation occurred over a wide front that encompassed several active island-shoal complexes. On the other hand, delta plain formation of the Brahmaputra occured inland along two loci created by channel avulsions east and west of the Pliestocene Madhupur terrace. The Sylhet basin, in the east of the Bengal delta, faced southward into

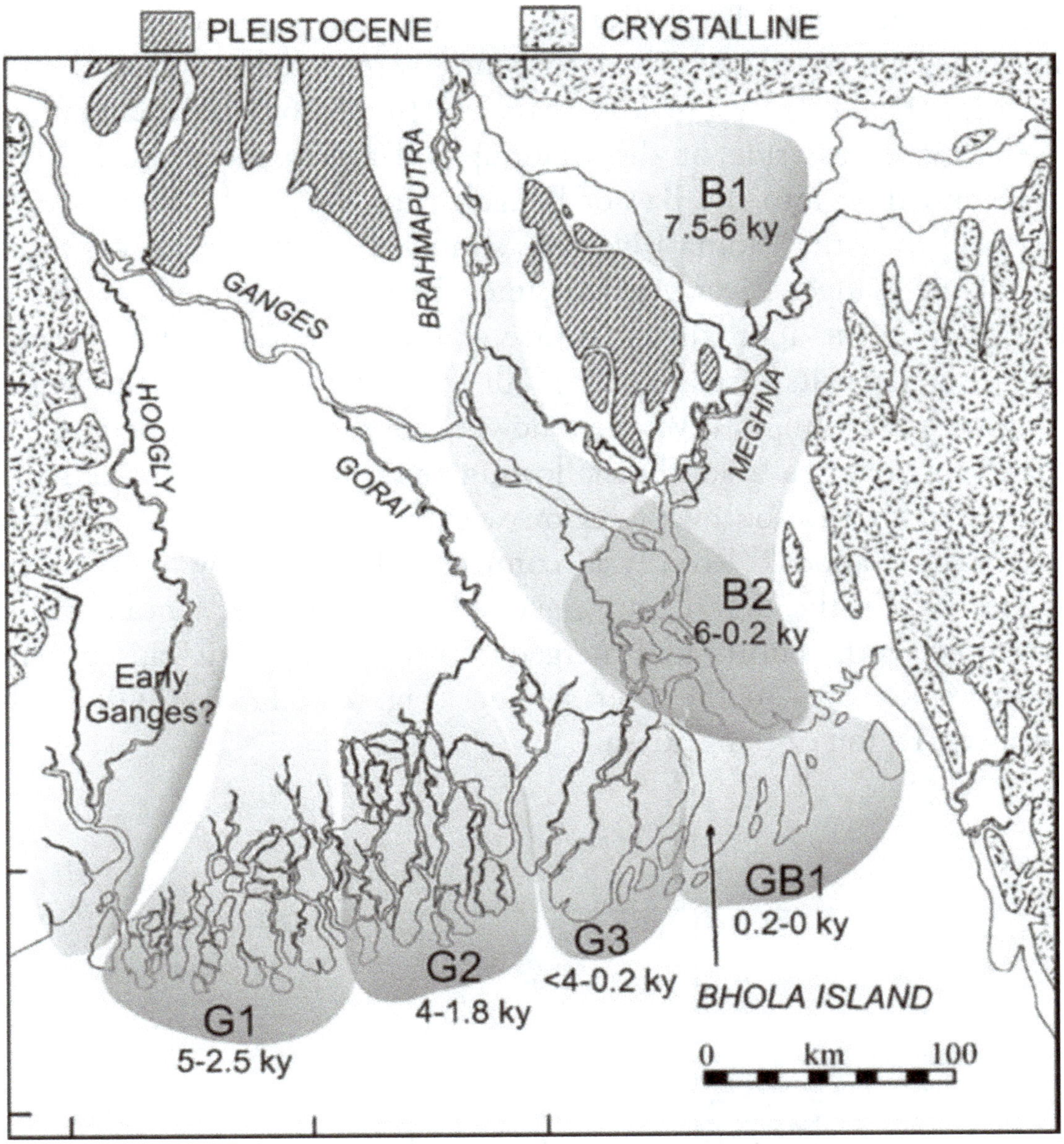

Fig. 2.2 Paleo-geographic map and the pathways and timing of the phases of late Holocene growth of the lower delta plain associated with the Ganges (G1, G2, G3), Brahmaputra (B1, B2) and combined Ganges-Brahmaputra (GB1) Deltas (Adapted from Sarker et al. 2013)

the Meghna estuary following the Meghna River course. Delta progradation into the Meghna estuary (GB1) was limited until the two rivers, the Ganges and the Brahmaputra, met in historical times. This progradation direction matches well with the findings of Goodbred and Kuehl (1998).

Consistent with the evolution of the river system during the Holocene, the rivers have changed course several times during the last centuries (Fig. 2.3). About 250 years ago (in 1776), the Brahmaputra River flowed east side of the Madhupur Tract to meet the Meghna River, and then into the Bay of Bengal, with the Bhola district on its west. Till then the Padma River was simply the downstream continuation of the Ganges River. At that time, the Ganges entered the Bay of Bengal along the approximate course of the Arial Khan River keeping Bhola island on its east (Fig. 2.3). All the distributaries in this region along with the Ganges River were flowing southeast in that period. The Chandana- Barasia, a southeast flowing river, was the main source of freshwater for the southwest region with a small link with the Gorai River. The Kabodak River was connected by a narrow link to the Ganges. When the Brahmaputra avulsed to the present Jamuna in the early nineteenth century and merged with the Ganges, it caused many significant changes to the river systems (Fig. 2.3, 1776–1840) in the southwest region of Bangladesh.

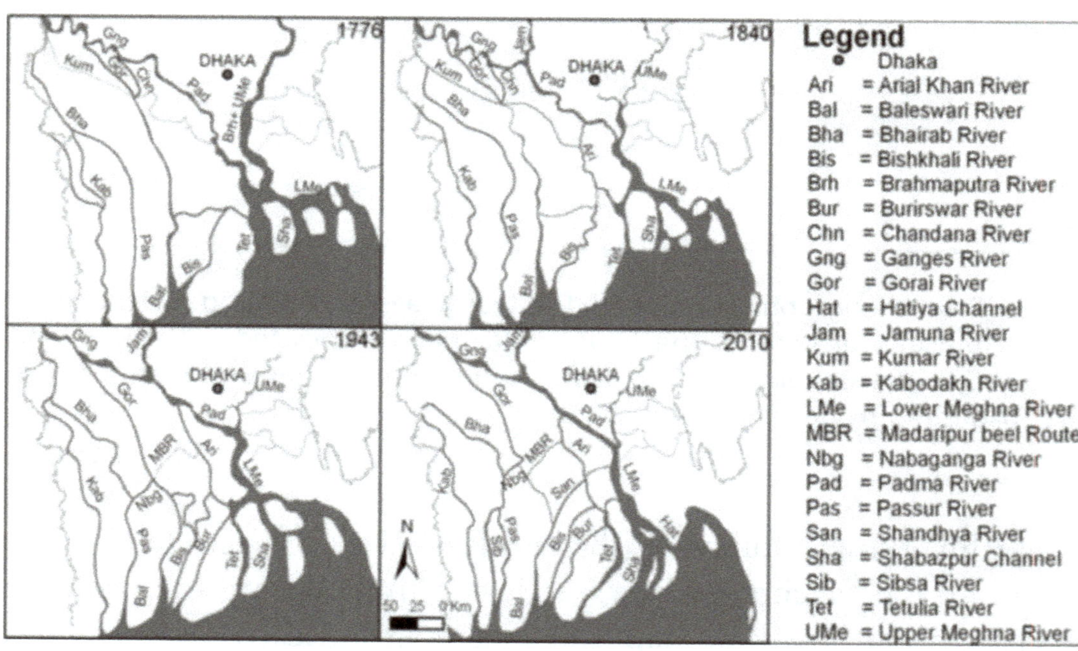

Fig. 2.3 Development of the main rivers in Bangladesh from 1776 to 2010 (Adapted from Sarker et al. 2013)

During the last 250 years, the location of the delta-building estuary has moved eastward. In the same period the distributaries, which also contribute in the delta-building processes, shifted their courses to the southwest direction, the dominating direction of which was mainly southeast about 200 years ago (Sarker et al. 2013).

Currently, the GBM Delta is fed and drained by the rivers Ganges, Brahmaputra and Meghna, along with their numerous tributaries and distributaries. The Ganges River drains the Himalayas and a significant portion of Northern India for approximately 2500 km before entering the Bengal basin and dividing into two distributaries. The main stem of the Ganges continues flowing southwards where it meets the Brahmaputra River. The other stem flows through West Bengal in India as the Bhagirathi-Hoogly River (Fig. 2.1). Cumulative river water discharge through the GBM Delta is the fourth largest in the world (Milliman and Meade 1983) and the delta is the world's largest sediment dispersal system (Kuehl et al. 1989). It has been estimated that about 10^3 million tonnes of sediment per year pass through to the Bay of Bengal across the 380 km delta front (Allison 1998), although analysis of recent sediment flux suggests a reduction by 50% with a decreasing trend (Rahman et al. 2018). The GBM Delta has a mean rate of subsidence of 3.9 mm/yr (Brown and Nicholls 2015). This translates into an average relative sea-level rise, including climate-induced rise, of about 7 mm/yr over the last few decades, although there is spatial variability.

2.3 Delta Development During the Anthropocene

The modern history of settlements in this region dates back to 1757 when land passed from local landowners to the East India Company, who began reclamation in the Sundarbans for rice fields through direct leasing of land to local farmers who enclosed sections of land. In 1839, rights for forest land with 99 years lease agreements known as '*Latdars*' or tenure holders were also issued, with land sales initiated in 1865. In the precolonial period most of the pasture lands of rural Bengal were under the control of the village community. However, during the colonial period (post 1757), pressure of population and permanent settlement generated changes. Land came under the control of revenue collectors known as *Zamindars*, *Jatdars*, and *Jaigirdars* who, with supportive state intervention, converted pasture land into cultivated land. Consequently, during this period the delta population remained reliant on mainly small-scale agricultural productivity (see Ghosh 2017).

Infrastructure Development

The advent of the Anthropocene (after 1950) coincides with Independence for India from the British Empire in 1947. In the post-colonial era, many large river dams were built across Indian rivers to facilitate either irrigation or power generation (Alley et al. 2014); interception of the rivers and construction of reservoirs being regarded as the most convenient method of water storage. One of the clear changes in the delta since this date is that more than five thousand dams have been built in the upper catchments, submerging extensive areas in the upper delta and, with them, the homelands of at least 40 million people. However, the benefits of projects, be it irrigation or hydropower, were generally enjoyed by the people living in the lower catchment. In addition, the delineation of the Indo-Bangladesh border resulted in 54 rivers including the Ganga becoming transboundary rivers, leading to conflicts around the sharing of water.

The commissioning of the Farakka Barrage on the Ganga in India in 1975 (Fig. 2.1) was expected to improve the status of the navigation channel approaching the port of Kolkata as the water diverted from the Ganga to the Bhagirathi was expected to reduce the sedimentation in the estuary and ensure better draught for the ships. However, this did not occur and sedimentation remains an unsolved problem (Rudra 2018). This was closely followed by the Teesta Barrage Project (TBP), conceived by the Irrigation and Waterways Department, Government of West Bengal in 1976, with the vision of irrigating around 1,000,000 ha of land. Seven hydropower stations, four in Sikkim and three in West Bengal, on the Teesta and its tributaries have been constructed to date. The fluctuation of flow has affected flora and fauna downstream (Rudra 2018). Construction of major engineering projects has continued. Following the Gazoldoba Barrage in India (1987), Bangladesh constructed a barrage at Duani (Lalmonirhat district) in 1990. As in the development of the TBP, the planners did not consider available water at the barrage site. Consequently, there remained a wide gap between the potential benefits and the area actually irrigated. As such, the project only provides supplementary irrigation to the Kharif crop during mid-monsoon breaks. These represent a few examples of how historical river interventions have influenced the hydrological regime and delta development of the GBM during the Anthropocene.

Population and Land Use

The Anthropocene has seen a rapid increase in population for the combined GBM Delta (Fig. 2.4) and changes in land character (see Figs. 2.5 and 2.6), although it remains one of the most under-developed areas in both Bangladesh and India. The population is mostly dependent on traditional monocrop (*Aman* paddy/rice) cultivation and riverine/offshore fishing, crab collection, honey collection, among other livelihoods. The recent decline in agricultural productivity, linked with poverty, is increasing the movement of migrants out of the delta for dominantly economic reasons (Adger et al. 2018; Hajra and Ghosh 2018). However, the percentage of urban population is increasing gradually (now around 27% of total) and this is contributing towards urbanising the delta, leading to changes in land use.

A key characteristic of the Anthropocene delta has been the progressive land use change driven by the introduction of large-scale aquaculture, partially for financial reasons and partially in response the progressive increase in salinity driven by rising sea-levels and hydrological mismanagement in the upper delta systems. The terrestrial forest and water-based ecosystem services play a major role in the livelihood of the people, and gradual depletion impacts the traditional farm-based economy. In the Indian part of the GBM Delta, the initial land conversion

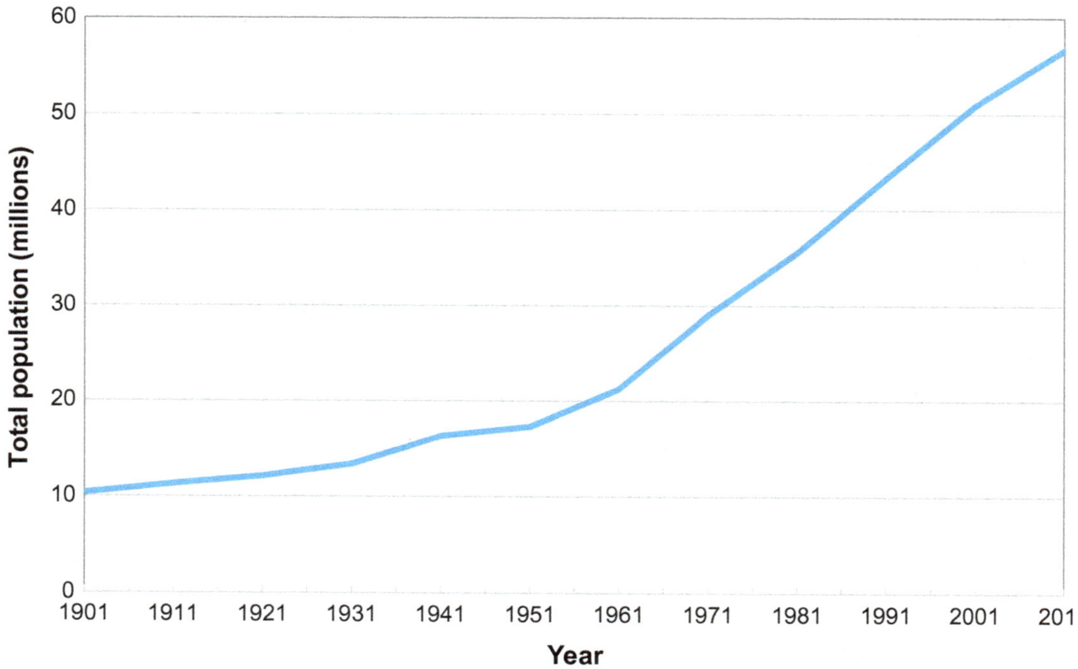

Fig. 2.4 Total population over time in the combined GBM Delta showing the rapid growth, post 1950, during the Anthropocene (Data from Census of India [2011], BBS [2015])

was from mangroves to agriculture and human settlement and, following a decline in agriculture, conversion to aquaculture and brickfields (Fig. 2.5). Saline water aquaculture also gained popularity because of increased water salinity and soil salinity. Because of the lack of leaching activities and continuous increase of salinity, productivity decreased gradually and land is no longer suitable for freshwater aquaculture. Finally, the land became only suitable for brickfields. This is one visible example of the impacts of unplanned land conversion in the Indian part of the GBM Delta. The Bangladesh part of the delta experienced a decline in agricultural area associated with a rapid increase in total built-up area and aquaculture as illustrated in Fig. 2.6. Aquaculture has grown in the southwest exploiting the saline water environment, with large economic returns, but at the expense of severe degradation of soil. There has been some increase in agricultural development in the southwest part of Bangladesh where improved drainage provides opportunities for new agriculture. Increases in planted mangrove areas and mudflats are

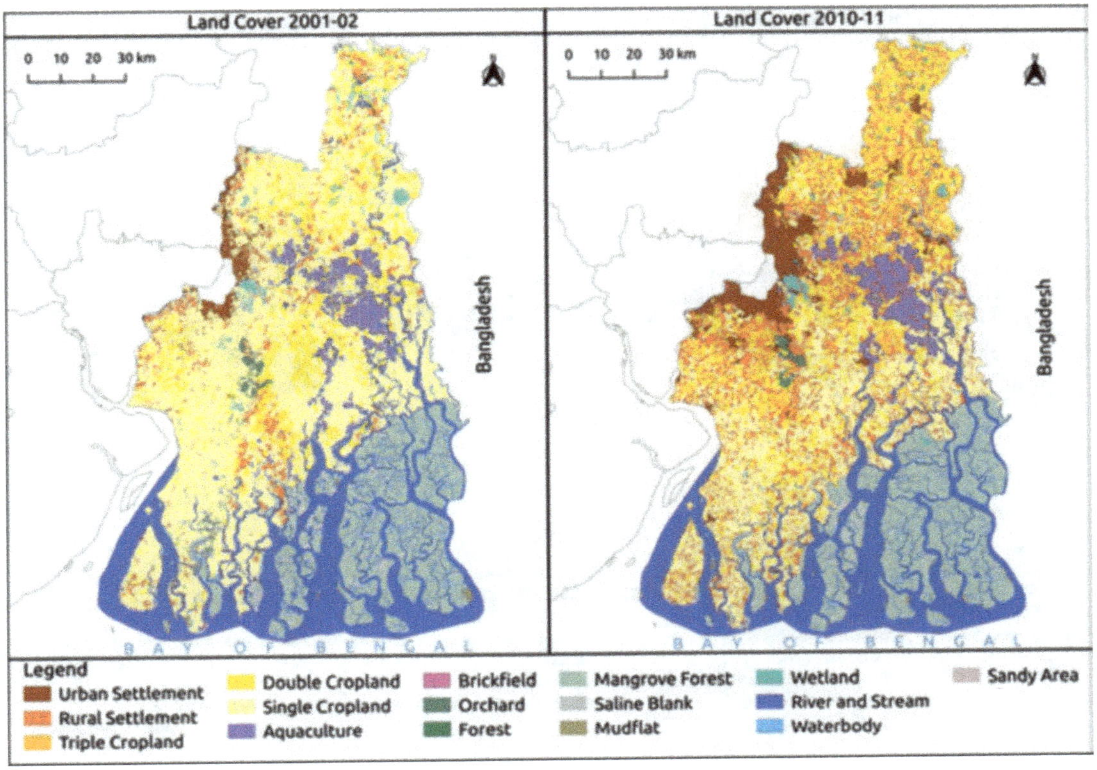

Fig. 2.5 Land use change in the Indian section of the GBM Delta showing the increase in urban settlements, decrease in agricultural land and area of land converted to brickfields

apparent, especially within Meghna Estuary. Increase in mudflat area is a manifestation of the active delta-building process, while increase in mangrove area on mudflats represents the increasing focus of the government on coastal afforestation. These new areas are being explored for industrial intensification through special economic zones, which target high growth and employment. The delta-building process in the GBM is thus opening further dimensions for human intervention in the delta during the Anthropocene, especially in Bangladesh.

Policy and Governance Interventions

India and Bangladesh have faced difficulties in cooperating to achieve policy formulation and management of the GBM Delta in the Anthropocene. The transboundary issues can be traced back to the formation of Bangladesh

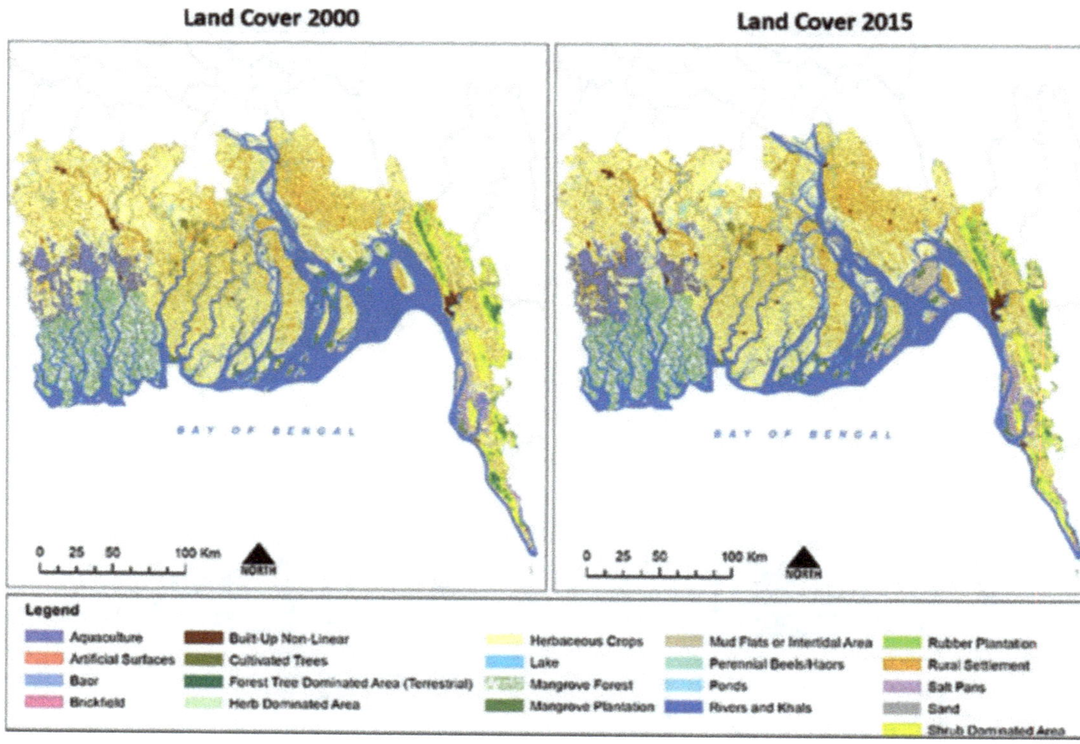

Fig. 2.6 Change in land use in the Bangladesh portion of the GBM Delta from 2000 to 2015 (see DECCMA 2018)

(1971) and, despite sharing fifty-four rivers with India, Bangladesh only signed the Ganges Treaty with India in 1996 with India removing the arbitration clauses with respect to minimum water requirements. Water sharing agreements should be implemented with comprehensive policy dialog, ensuring the principles of equity and fairness. The formation of the Indo-Bangladesh Joint River Commission (JRC) has not improved the situation, with blockage or diversion of river water not discussed between the two countries. For example, the construction of the Farakka Barrage in 1975 led to a sharp decline of the freshwater supply to the Ganges River during the dry season, adversely affecting agriculture, navigation, irrigation, fisheries and allowing salinity intrusion within Bangladesh.

Governments, non-governmental bodies and foreign organisations have been identifying innovative approaches to help the delta populations adapt to change and to reduce vulnerabilities in the face of future uncertainties. National Adaptation Programme of Action 2005, Bangladesh Climate Change Strategy and Action Plan 2009 (BCCSAP 2009) and

Bangladesh Climate Change and Gender Action Plan 2013 (BCCGAP 2013) are some of the guiding documents for planned adaptation strategies formulated by the Government of Bangladesh (Haq et al. 2015). Various climate change adaptation issues have been addressed to different extents in sectoral policies of Bangladesh with regards to dimensions such as risk reduction, community- and ecosystem-based adaptation, migration and gender. The National Action Plan on Climate Change (NAPCC) of 2008 and 2014 and the West Bengal State Action Plan on Climate Change (WBSAPCC) of 2012 have been the guiding documents in the Indian part of the delta. A total of seven (agriculture, wind, energy, health, waste to energy, health and coastal areas) out of twelve missions stated in the NAPCC promote adaptation, while the WBSAPCC marks the first integrated plan to combat climate change phenomenon in the Indian part of the delta (Dey et al. 2016). In both parts of the delta, the sectoral policies and plans have tended to be short term and sector-specific, following the mandate of the formulating ministries or departments, while clearly lacking integration across relevant issues including the cross-cutting issues. There is a clear need for a framework to allow coordination among the sectoral approaches from climate change perspective by setting sectoral priorities and identifying key sectors for immediate attention (Haq et al. 2015; Dey et al. 2016; Salehin et al. 2018).

In Bangladesh, awareness of gender emerged in policies and plans most notably after 2009. Although female empowerment is emphasised in areas such as disaster preparedness and management activities, agriculture management and agriculture wages, there is no explicit link between gender-specific needs in the climate change context. There is a lack of guidelines for gender-specific adaptation; policies and plans which address gender issues discuss 'what to do' but not 'how to do'. BCCGAP 2013 is the only dedicated document in this regard (Haq et al. 2015; Salehin et al. 2018). Sectoral policies in India have put major emphasis on women's role in agriculture, management of natural resources and empowerment of women with appropriate skill development. They have addressed to some extent women-targeted initiatives with regard to self-employment and small entrepreneurship. Gender issues need mainstreaming in sectoral policies, with a more gender sensitive and inclusive approach, from climate change and disaster management perspective.

As challenges to attaining national-level goals and climate change targets present significant downside risks and uncertainties which warrant long-term strategies, the government of Bangladesh has developed the Bangladesh Delta Plan 2100 (BDP2100 2018). This aims to integrate sectoral, national and global targets and plans into long-term coherent strategies across government ministries for adaptive delta management taking scenarios of climate change, population growth and economic development into account (BDP2100 2018). The focus is on steering the opportunities and vulnerabilities created by the interface of water, climate change, natural disasters, poverty and environment, with sustainable use of water resources and prevention of water-related natural disasters providing the backbone.

2.4 Adaptation, Migration and a Way Forward

The future of the GBM Delta in the Anthropocene has many opportunities, including tremendous potential for economic growth. At the same time, environmental hazards, particularly in so far as impacts are more keenly felt among the vulnerable populations, are a concern. Mitigating hazards offers relief from one side of the equation only; it does not constitute vulnerability reduction. This situation will be exacerbated by climate change during the Anthropocene and presents the countries in the GBM Delta with a variety of significant challenges. The impacts of climate change are expected to disrupt the complex hydrological balance existing in the GBM Delta basin and lead to a range of water management challenges. Factors such as salinity intrusion into rivers have already led to reduction of cropland and an ongoing transition to large-scale aquaculture (Amoako-Johnson et al. 2016). The much lower employment in aquaculture than agriculture leads to migration.

Exacerbating environmental pressures due to global climate and environmental change could create social destabilisation. Social vulnerability is defined as the inability of people, organisations and societies to withstand adverse impacts from the multiple stressors to which they are exposed (Adger 1999; Adger and Kelly 1999). Vulnerability and related characteristics have been characterised for deltas (Chapter 6), including coastal Bangladesh (Uddin et al. 2019). Here a first preliminary transboundary assessment of social vulnerability to adverse change is shown for the whole GBM Delta in Fig. 2.7. The analysis is based on Principal Component Analysis of 13 socio-economic variables explained with the rationale in Table 2.1. The data is derived from census data at the sub-district level (*Upazila* in Bangladesh, and Community Development Block in India) as follows the Census of India and the Bangladesh Bureau of Statistics for 2001 and 2011. The results show that there is a strong vulnerability gradient across the GBM Delta coast in both India and Bangladesh, with the highest vulnerability being closer to the Bay of Bengal. Analysis of the change in social vulnerability from 2001 to 2011 suggests that planned efforts to address poverty, generate non-farm employment and improve health and sanitation status have reduced social vulnerability, while climatic hazards such as the major cyclones of Sidr (2007) and Aila (2009) have increased social vulnerability in

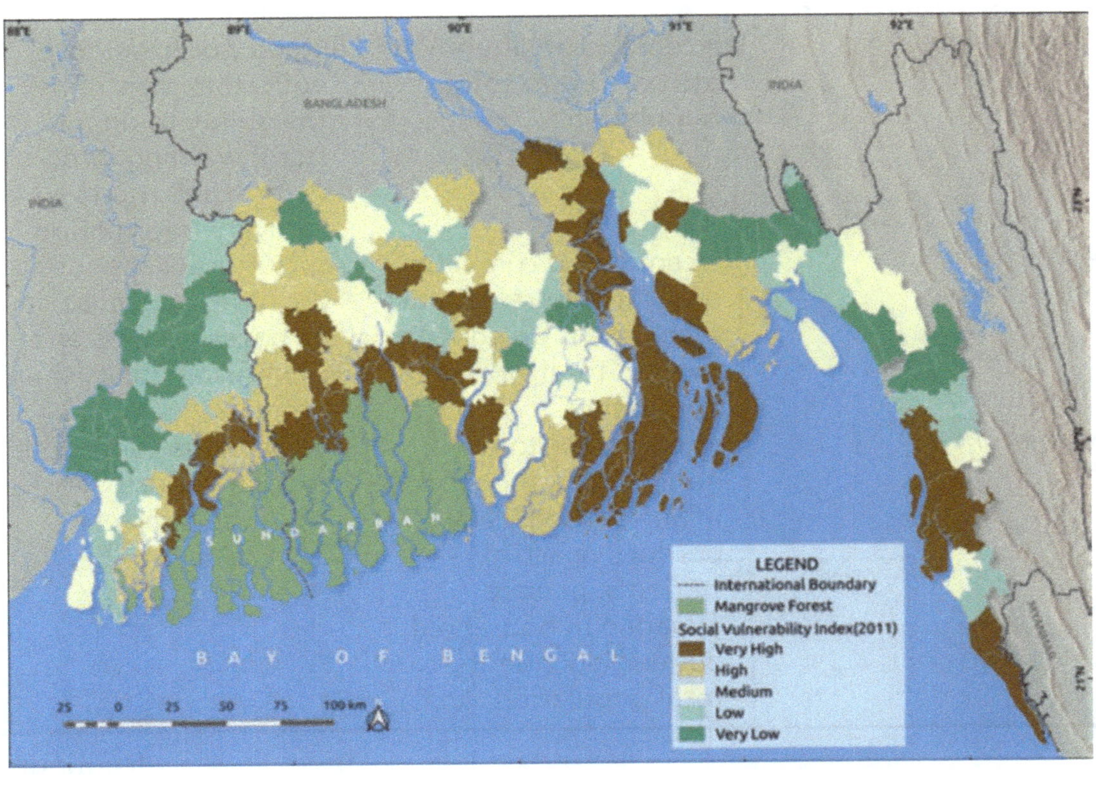

Fig. 2.7 Social Vulnerability Index across the coastal GBM Delta for the year 2011 (see Hazra and Islam 2017; also DECCMA 2018)

Table 2.1 Description of the socio-economic variables considered in the social vulnerability analysis of the GBM Delta

Variables		Rationale
Population density	Number of people per square kilometre	Areas with high population density are more exposed to climate change impacts
Average household size	Average number of people per household	Families with a large number of people have more limited resources and more work responsibilities that reduce the resilience to and recovery from hazards
Female population	Percentage of female population to total population	Females have a more difficult time during recovery from disasters than males, due to their family care responsibilities, sector-specific employment and lower wages
Illiteracy rate	Percentage of illiterate persons to total population	Illiteracy or lower level education constrains the ability to understand warning information and access to recovery information
Agricultural dependency	Percentage of cultivators and agricultural labours (dependent on agriculture) to total working population	Agricultural dependents are more impacted by hazard events and climate variability than other workers
Non-workers	Percentage of total non-workers (no work in any economically productive activity—students, persons engaged in household duties, dependents) to total population	Non-workers contribute to a slower recovery from the disasters

(continued)

Table 2.1 (continued)

Variables		Rationale
Kutcha house	Percentage of households living in Kutcha (walls and/or roof are predominantly made by mud, bamboos, grass, reeds, thatch, plastic/polythene) houses (temporary structure)	People living in Kutcha houses are more vulnerable to hazards
Home ownership	Percentage of households that do not own their home (rented, occupied and others)	People who don't own their home have less access to information about financial aid during recovery
Electricity connection	Percentage of households living without mains electricity connection	Households without access to safe/improved sources of drinking water, electricity connection and sanitation facility are more sensitive to climate change impacts. They have the lower ability to respond to and recover from the impacts of hazards
Drinking water	Percentage of households reported 'others' category (i.e. ponds/canal/spring/river) as the main source of drinking water	
Sanitation facility	Percentage of households that have no sanitation facility	
Poverty	Percentage of population living below the poverty line	Poor people have lower access to resources and lower ability to absorb losses and enhance resilience to hazard impacts
Rural population	Percentage of population living in rural areas (total population minus urban population)	Rural populations are more dependent on natural resources and have lower incomes

affected sub-districts. Similar issues exist across the GBM Delta in India and Bangladesh, and similar solutions are appropriate.

The delta has limited capacity to deal with current climate and hydrological variability and, without major capacity development, intensification expected from climate change will exacerbate problems as the Anthropocene unfolds. The Intergovernmental Panel on Climate Change (IPCC) Fourth Assessment Report (AR4) highlighted South Asia and the greater Himalaya as one of several key regions having greatly divergent predictions of future changes in precipitation. Both climatic and non-climatic events adversely affect the livelihood of the people of this delta. Land loss due to submergence and increasing soil salinity, along with land fragmentation, will result in challenges for the delta livelihoods.

In the GBM Delta, economic reasons dominate people's perception as the important drivers of migration (Arto et al. 2019). However, environmental reasons are also viewed as a component driver; albeit by a smaller percentage of respondents (about 1.5% people identified environmental degradation or extreme events as the first most important driver, another 5.3% as the second most important driver and another 10% as the third most important driver). In the GBM, West Bengal, almost two-thirds of the migrants are moving to seek better employment, followed by family obligations (12.3%), while 10% left their origin to pursue a degree or obtain training in a new skill. Only 3% of the population cited environmental stresses as the direct cause of migration (see Chapter 7). The fact that environmental stresses often precipitate economic stresses also suggests that people do not always clearly perceive the causes of economic stresses. Environmental factors might be playing a bigger role than the numbers suggest. Intentions to migrate in the future are high (among two-thirds of all households), with seeking jobs, better education and environmental stresses as important reasons. Perceived environmental impacts (e.g. flooding, cyclone, erosion), including loss of seasonal income, are apparent in areas more exposed to hazards, indicating a higher probability of future migration from the more hazard-prone areas (see Chapter 7). In policies and plans, migration has not been addressed as a climate change adaptation option, either in Bangladesh or India. In Bangladesh, the emphasis has been on overseas migration, while rural to urban migration has been discouraged. This resulted in adaptation gaps in moulding internal migration into economic benefits. Recent emphasis has been on fostering economic growth and employment opportunities in coastal areas through labour-intensive industries in planned industrial zones.

Adaptation policies in both India and Bangladesh are mostly disaster focused, lacking sectoral coherence and with little focus on gender-specific adaptation and migration as a climate change adaptation (see Chapter 9). Adaptation policy analysis in Bangladesh has revealed that climate change issues are mostly disaster risk reduction (DRR) focused, with ecosystem-based adaptations and community-based adaptations recently emerging (Haq et al. 2015). Most adaptation measures are infrastructure focused, in alignment with earlier policies, and reactive in nature, with inadequate focus on gender issues. Research has revealed an inventory of implemented adaptations showing that DRR has been the major focus in almost one-fourth of the total adaptations in the country, followed by water resources management (WRM) (20%), infrastructural development (17%) and agriculture (13%). The coastal districts (the delta region), being the most vulnerable region to natural hazards and disasters has received more focus on DRR (73% of total DRR adaptations) and WRM (60% of total WRM adaptations) (Haq et al. 2015). Most of the WRM, DRR and coastal zone management focused adaptations to climate variability and climate change are infrastructural in nature which suggests a priority of implementing organisations (mostly government) towards infrastructural development. Even with the commendable task of preparing BCCSAP 2009 as the first LDC country and having Climate Change Trust Fund (CCTF) in 2010, there has been relatively little emphasis on the capacity building which was a key aspiration in BCCSAP 2009 to better understand climate change risks and improve planning and execution in development projects to combat climate change. Most of the adaptation measures have been implemented in recent decades when climate change manifestations have been clear, though some of the development activities occurring in the twentieth century have served in combating climate change induced disastrous events. Nevertheless, adaptation activities with anticipation of future major hazards have been less in number while the majority have been reactive in nature. This explains why the majority of the adaptations (75% in the whole country and 72% in the coastal delta) are undertaken in response to chronic stresses like salinity and waterlogging in the delta region and river bank erosion, regular flood and drought in the non-delta region, rather than sudden shocks like major cyclonic storm surge events and large floods.

The infrastructure focused adaptation initiatives have meant a skewed allocation of adaptation funds across the implementing ministries. Lack of guidance on gender-specific adaptation needs in policies and plans and lack of participation from local people resulted in less than one-fourth of adaptation addressing gender issues. Examination of the projects executed under the Bangladesh CCTF showed that projects executed by the local government institutions (LGIs) were more targeted towards climate change compared to the central government agencies targeting climate variability induced hazards. Gender-sensitive issues were also considered better in adaptation projects implemented by LGIs than projects executed by central Government of Bangladesh agencies. More inclusive planning involving LGIs and NGOs may improve gender-sensitive development and thus overall sustainability of adaptation options.

Agricultural livelihoods, rural development and DRR are the major sectors for adaptation activities in West Bengal. Most of the activities are reactive approaches in response to Cyclone Aila of 2009. More anticipatory approaches such as setting up a Climate Change Adaptation Centre, renewable energy, livelihood development of disadvantaged sections, seed banks, mobile boat dispensary, probable human migration, innovative farming practices are yet to be explored. The majority of the adaptation activities have been undertaken due to 'stress' in the area, while only few have been due to 'shock'. The factors of changing climatic conditions, cyclones and storm surges, breaching of embankments leading to coastal inundation, floods, coastal erosion have been pointed out as the main cause of stress. Apart from this, the other triggers of adaptation include loss of livelihood and income, unemployment, loss of physical assets and lack of proper market linkages. In terms of adaptation providers, the majority of reported adaptations have been provided by the government.

Bangladesh requires a long-term vision, planning and implementation comprising all government ministries and agencies that contribute to this collective objective. Owing to the large uncertainties with regard to climate change and socio-economic progress, planning, robust and versatile strategies are required for effective Adaptive Delta Management (Seijger et al. 2017; BDP2100 2018). Policymakers from both India and Bangladesh require stronger knowledge and scientific tools to anticipate the dynamic effects of global climate change and their interaction with environmental and socio-economic change and make decisions on the most appropriate interventions and investments (e.g. Nicholls et al. 2018).

Current adaptation practices in the GBM are shown in Fig. 2.8. These vary widely from east to west which are fundamentally triggered by the adaption need at a local environment level that includes stresses due to disasters such as flooding, erosion, salinisation, storm surge, etc. Cross-border learning on effective adaptation practices that require collaborative research on joint GBM issues will promote replicable future adaptations. The provision of effective cyclone warning systems and

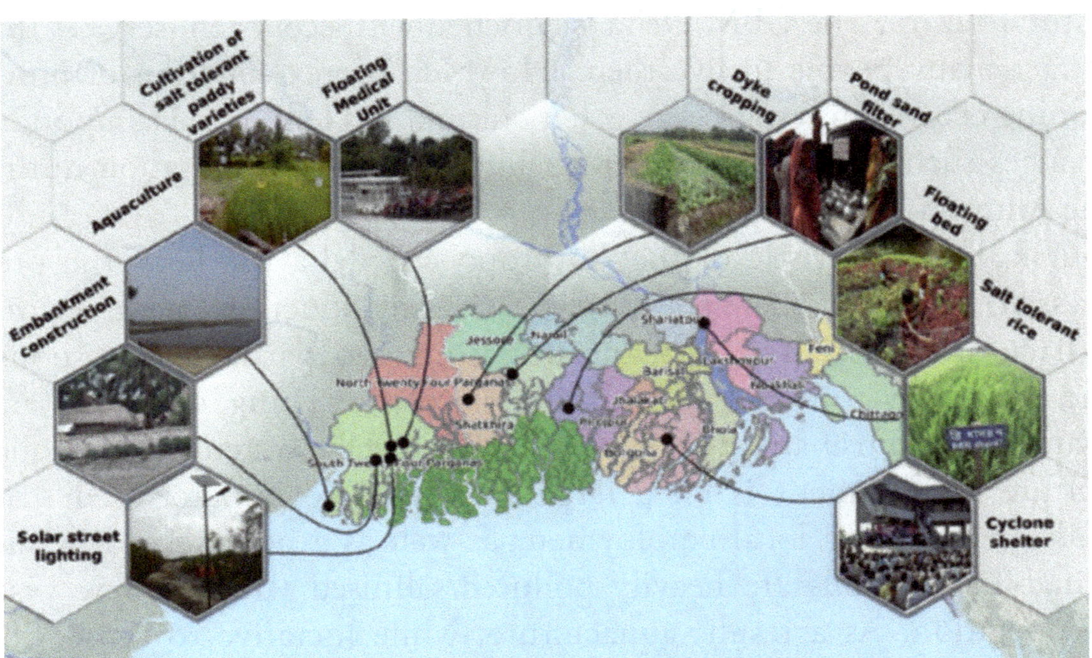

Fig. 2.8 Examples of adaptation options utilised in the GBM Delta

thousands of cyclone shelters provides one strong example of recent adaptation success in Bangladesh and West Bengal (Lumbroso et al. 2017). Looking to the future, strategic development of tidal river management towards more systematic controlled sedimentation within polders offers the promise of a new innovative response to relative sea-level rise (discussed in more detail in Chapters 5, 6, 11).

2.5 Conclusions

Coastal regions of the GBM Delta have long been settled by humans due to their abundant resources for livelihoods, including agriculture, fisheries, transportation and rich biodiversity. However, natural and anthropogenic factors, such as climate change and sea-level rise, and land subsidence, population pressure and developmental activities during the Anthropocene pose threats to the delta's sustainability. Indeed, there is a strong association between household poverty and the likelihood of material and human loss following a natural event in the islands of the Indian GBM. It is also evident that the poorest households are most likely to suffer from deteriorating livelihoods following a natural hazard. The GBM Delta is enriched with ecosystem services that attract many people to live there. However, overexploitation of provisioning ecosystem resources (e.g. agriculture) has led to other important changes such as declining water quality and higher salinisation during the Anthropocene.

A key characteristic of the Anthropocene delta has been the progressive land use change driven by the introduction of large-scale aquaculture, partially for financial reasons and partially in response the progressive increase in salinity driven by rising sea-levels and hydrological mismanagement in the upper delta systems. This aquaculture development is on a large and highly commercialised scale and offers limited local employment as well as a poor record in sustainability, as soils are heavily polluted/salinised within a few years of operation. As a result, aquaculture, while lucrative to those who invest in it, contributes to the loss of livelihoods of the poorest sectors of the community and consequently migration out of the region.

This supports an ongoing migratory trend, predominantly due to economic drivers, from the delta.

The growing populations of the two associated delta megacities (Dhaka and Kolkata) indicate continuous in-migration in recent years. This produces a population that is highly vulnerable within the urban environment, subject to poor working and wage conditions and highly susceptible to urban climatic change. Making these two cities sustainable in the long run requires greater policy recognition of the drivers of migration and their fundamentally economic underpinnings, and more specifically that people will continue to migrate to access urban benefits. Indeed, it is perhaps better to undertake a strategy of bringing the urban benefits to those in rural areas, such as improved infrastructure and development of smaller distributed cities and the health, economic, livelihood and education benefits that entails than simply aiming to enhance standard rural livelihoods. As such, in the Anthropocene it is vital that the delta is understood and developed as a system, considering both the urban and rural areas, which will feed the cities and sustain biodiversity, and all the threats and opportunities are considered. This is consistent with evidence-based Adaptive Delta Planning as proposed in the BDP2100 (2018).

References

Adger, W. N. (1999). Social vulnerability to climate change and extremes in coastal Vietnam. *World Development, 27*(2), 249–269. https://doi.org/10.1016/S0305-750X(98)00136-3.

Adger, W. N., Adams, H., Kay, S., Nicholls, R. J., Hutton, C. W., Hanson, S. E., et al. (2018). Ecosystem services, well-being and deltas: Current knowledge and understanding. In R. J. Nicholls, C. W. Hutton, W. N. Adger, S. E. Hanson, M. M. Rahman, & M. Salehin (Eds.), *Ecosystem services for well-being in deltas: Integrated assessment for policy analysis* (pp. 3–27). Cham: Springer. https://doi.org/10.1007/978-3-319-71093-8_1.

Adger, W. N., & Kelly, P. M. (1999). Social vulnerability to climate change and the architecture of entitlements. *Mitigation and Adaptation Strategies for Global Change, 4*(3), 253–266. https://doi.org/10.1023/A:1009601904210.

Alam, M. (1996). Subsidence of the Ganges—Brahmaputra Delta of Bangladesh and associated drainage, sedimentation and salinity problems. In J. D. Milliman & B. U. Haq (Eds.), *Sea-level rise and coastal subsidence: Causes, consequences, and strategies* (pp 169–192). Dordrecht, The Netherlands: Springer. https://doi.org/10.1007/978-94-015-8719-8_9.

Alley, K. D., Hile, R., & Mitra, C. (2014). Visualizing hydropower across the Himalayas: Mapping in a time of regulatory decline. *Himalaya, the Journal of the Association for Nepal and Himalayan Studies, 34*(2), 9.

Allison, M. A. (1998). Historical changes in the Ganges-Brahmaputra Delta front. *Journal of Coastal Research, 14*(4), 1269–1275.

Allison, M. A., Khan, S. R., Goodbred, S. L., & Kuehl, S. A. (2003). Stratigraphic evolution of the late Holocene Ganges-Brahmaputra lower delta plain. *Sedimentary Geology, 155*(3), 317–342. https://doi.org/10.1016/S0037-0738(02)00185-9.

Amoako-Johnson, F., Hutton, C. W., Hornby, D., Lázár, A. N., & Mukhopadhyay, A. (2016). Is shrimp farming a successful adaptation to salinity intrusion? A geospatial associative analysis of poverty in the populous Ganges–Brahmaputra–Meghna Delta of Bangladesh. *Sustainability Science, 11*(3), 423–439. https://doi.org/10.1007/s11625-016-0356-6.

Arto, I., García-Muros, X., Cazcarro, I., González, M., Markandya, A., & Hazra, S. (2019). The socioeconomic future of deltas in a changing environment. *Science of the Total Environment, 648,* 1284–1296. https://doi.org/10.1016/j.scitotenv.2018.08.139.

BBS. (2015). *Population density and vulnerability: A challenge for sustainable development of Bangladesh.* Population Monograph No. 7. Bangladesh Bureau of Statistics (BBS), Ministry of Planning, Government of the People's Republic of Bangladesh. http://203.112.218.65:8008/PageWebMenuContent.aspx?MenuKey=243. Last accessed 5 December 2018.

BDP2100. (2018). *Bangladesh Delta Plan 2100.* Volumes 1-Strategy and 2-Investment Plan. Dhaka, Bangladesh, General Economics Division (GED), Bangladesh Planning Commission, Government of the People's Republic of Bangladesh. https://www.bangladeshdeltaplan2100.org/. Last accessed 8 October 2018.

Biswas, A. K. (1992). Indus water treaty: The negotiating process. *Water International, 17*(4), 201–209. https://doi.org/10.1080/02508069208686140.

Brown, S., & Nicholls, R. (2015). Subsidence and human influences in mega deltas: The case of the Ganges–Brahmaputra–Meghna. *Science of the Total Environment, 527,* 362–374. https://doi.org/10.1016/j.scitotenv.2015.04.124.

Census of India. (2011). *Table A2 area, population, decennial growth rate and density for 2001 and 2011 at a glance for West Bengal and the districts*. Office of the Registrar General and Census Commissioner, Government of India. http://censusindia.gov.in/2011-prov-results/prov_data_products_wb.html. Last accessed 5 November 2018.

Dastagir, M. R. (2015). Modeling recent climate change induced extreme events in Bangladesh: A review. *Weather and Climate Extremes, 7*, 49–60. https://doi.org/10.1016/j.wace.2014.10.003.

DECCMA. (2018). *Climate change, migration and adaptation in deltas: Key findings from the DECCMA project* (Deltas, Vulnerability and Climate Change: Migration and Adaptation [DECCMA] Report). Southampton, UK: DECCMA Consortium. https://www.preventionweb.net/publications/view/61576. Last accessed 27 November 2018.

Dey, S., Ghosh, A. K., & Hazra, S. (2016). *Review of West Bengal State adaptation policies, Indian Bengal Delta* (Deltas, Vulnerability and Climate Change: Migration and Adaptation [DECCMA] Working Paper). Southampton, UK: DECCMA Consortium. https://generic.wordpress.soton.ac.uk/deccma/resources/working-papers/. Last accessed 6 August 2018.

Ericson, J. P., Vörösmarty, C. J., Dingman, S. L., Ward, L. G., & Meybeck, M. (2006). Effective sea-level rise and deltas: Causes of change and human dimension implications. *Global and Planetary Change, 50*(1–2), 63–82. https://doi.org/10.1016/j.gloplacha.2005.07.004.

Ghosh, S. (2017). Land reforms, agitations, tensions in Malda and West Dinajpur: Tebhaga to Naxalbari. *International Journal for Innovative Research in Multidisciplinary Field, 3*(5), 163–168.

Goodbred, S. L., & Kuehl, S. A. (1998). Floodplain processes in the Bengal Basin and the storage of Ganges-Brahmaputra river sediment: An accretion study using 137Cs and 210Pb geochronology. *Sedimentary Geology, 121*(3), 239–258. https://doi.org/10.1016/S0037-0738(98)00082-7.

Gopal, B., & Chauhan, M. (2006). Biodiversity and its conservation in the Sundarban Mangrove Ecosystem. *Aquatic Sciences, 68*(3), 338–354. https://doi.org/10.1007/s00027-006-0868-8.

Hajra, R., & Ghosh, T. (2018). Agricultural productivity, household poverty and migration in the Indian Sundarban Delta. *Elementa Science of the Anthropocene, 6*(1), 3. https://doi.org/10.1525/elementa.196.

Haq, M. I., Omar, M. A. T., Zahra, Q. A., Shashi, I. J., & Rahman, M. R. (2015). *Evaluation of adaptation policies in GBM Delta of Bangladesh* (Deltas, Vulnerability and Climate Change: Migration and Adaptation [DECCMA] Working Paper). Southampton, UK: DECCMA Consortium.

https://generic.wordpress.soton.ac.uk/deccma/resources/working-papers/. Last accessed 19 November 2018.

Haque, A., & Nicholls, R. J. (2018). Floods and the Ganges-Brahmaputra-Meghna Delta. In R. J. Nicholls, C. W. Hutton, W. N. Adger, S. E. Hanson, M. M. Rahman, & M. Salehin (Eds.), *Ecosystem services for well-being in deltas: Integrated assessment for policy analysis* (pp. 147–159). Cham: Springer. https://doi.org/10.1007/978-3-319-71093-8_8.

Hazra, S., & Islam, N. (2017, April 23–28). *A preliminary assessment of social vulnerability in Ganga-Brahmaputra-Meghna Delta*. EGU General Assembly, Vienna, Austria. European Geosciences Union.

Islam, A. S., Haque, A., & Bala, S. K. (2010). Hydrologic characteristics of floods in Ganges–Brahmaputra–Meghna (GBM) Delta. *Natural Hazards, 54*(3), 797–811. https://doi.org/10.1007/s11069-010-9504-y.

Kuehl, S. A., Hariu, T. M., & Moore, W. S. (1989). Shelf sedimentation off the Ganges-Brahmaputra river system: Evidence for sediment bypassing to the Bengal fan. *Geology, 17*(12), 1132–1135.

Lumbroso, D. M., Suckall, N. R., Nicholls, R. J., & White, K. D. (2017). Enhancing resilience to coastal flooding from severe storms in the USA: International lessons. *Natural Hazards and Earth System Sciences, 17*(8), 1357–1373. https://doi.org/10.5194/nhess-17-1357-2017.

Milliman, J. D., & Meade, R. H. (1983). World-wide delivery of river sediment to the oceans. *The Journal of Geology, 91*(1), 1–21. https://doi.org/10.1086/628741.

Morgan, J. P., & McIntire, W. G. (1959). Quaternary geology of the Bengal basin, East Pakistan and India. *Geological Society of America Bulletin, 70*(3), 319–342.

Mortreux, C., Safra de Campos, R., Adger, W. N., Ghosh, T., Das, S., Adams, H., et al. (2018). Political economy of planned relocation: A model of action and inaction in government responses. *Global Environmental Change, 50,* 123–132. https://doi.org/10.1016/j.gloenvcha.2018.03.008.

Nicholls, R. J., Hutton, C., Adger, W. N., Hanson, S. E., Rahman, M. M., & Salehin, M. (Eds.). (2018). *Ecosystem services for well-being in deltas: Integrated assessment for policy analysis*. London, UK: Palgrave Macmillan.

Noor, S. (2018). *Investigation of polderization induced water logging and feasible adaptation measures in Dumuria Upazila under Khulna Districts* (Unpublished Masters thesis). Institute of Water and Flood Management, BUET, Dhaka, Bangladesh.

Rahman, M., Dustegir, M., Karim, R., Haque, A., Nicholls, R. J., Darby, S. E., et al. (2018). Recent sediment flux to the Ganges-Brahmaputra-Meghna

Delta system. *Science of the Total Environment, 643,* 1054–1064. https://doi.org/10.1016/j.scitotenv.2018.06.147.

Raman, K. S., Kumar, S., & Neogi, B. B. (1986). *Exploration in Bengal Basin India—An overview.* Singapore: Offshore South East Asia Show. Society of Petroleum Engineers. https://doi.org/10.2118/14598-MS.

Rudra, K. (2018). Conflicts over sharing the waters of transboundary rivers. In *Rivers of the Ganga-Brahmaputra-Meghna Delta* (pp. 163–172). Cham: Springer.

Salehin, M., Rahman, R., Allan, A., Hossen, M. A., Chowdhury, A., & Sayan, C. (2018). *Challenges of governance system in addressing climate change adaptation measures in Bangladesh: Gaps, strengths and opportunities* (Deltas, Vulnerability and Climate Change: Migration and Adaptation [DECCMA] Policy Brief). Southampton, UK: DECCMA Consortium. http://generic.wordpress.soton.ac.uk/deccma/resources/briefs/. Last accessed 7 January 2019.

Sarker, M. H., Akter, J., & Rahman, M. M. (2013). Century-scale dynamics of the Bengal delta and future development. In *Proceedings of the International Conference on Water and Flood Management* (pp. 91–104). Dhaka, Bangladesh. https://edepot.wur.nl/317989. Last accessed 20 August 2018.

Seijger, C., Douven, W., van Halsema, G., Hermans, L., Evers, J., Phi, H. L., et al. (2017). An analytical framework for strategic delta planning: negotiating consent for long-term sustainable delta development. *Journal of Environmental Planning and Management, 60*(8), 1485–1509. https://doi.org/10.1080/09640568.2016.1231667.

Uddin, M. N., Saiful Islam, A. K. M., Bala, S. K., Islam, G. M. T., Adhikary, S., Saha, D., et al. (2019). Mapping of climate vulnerability of the coastal region of Bangladesh using principal component analysis. *Applied Geography, 102,* 47–57. https://doi.org/10.1016/j.apgeog.2018.12.011.

Woodroffe, C. N., Nicholls, R. J., Saito, Y., Chen, Z., & Goodbred, S. L. (2006). Landscape variability and the response of Asian megadeltas to environmental change. In N. Harvey (Ed.), *Global change and integrated coastal management: The Asia-Pacific region* (pp. 277–314). New York, NY: Springer.

3

The Mahanadi Delta: A Rapidly Developing Delta in India

Sugata Hazra, Shouvik Das, Amit Ghosh, Pokkuluri Venkat Raju and Amrita Patel

3.1 The Mahanadi Delta

The Mahanadi Delta (Fig. 3.1), located in the state of Odisha on the east coast of India, is a composite delta fed by water, sediments and nutrients from a network of three major rivers: the Mahanadi River (and its distributaries; the Devi, Daya, Bhargavi, Kushbhandra and Parchi) and the adjoining Brahmani and Baitarini Rivers (Kumar and Bhattacharya 2003). The 851 km long Mahanadi River has one of the largest drainage basin on the east coast of India (Fig. 3.1 inset)

S. Hazra (✉) · S. Das · A. Ghosh
School of Oceanographic Studies,
Jadavpur University, Kolkata, India
e-mail: sugata.hazra@jadavpuruniversity.in

P. V. Raju
National Remote Sensing Center,
Indian Space Research Organisation, Hyderabad, India

A. Patel
Sansristi, Bhubaneswar, Odisha, India

R. J. Nicholls et al. (eds.), *Deltas in the Anthropocene*,
https://doi.org/10.1007/978-3-030-23517-8_3

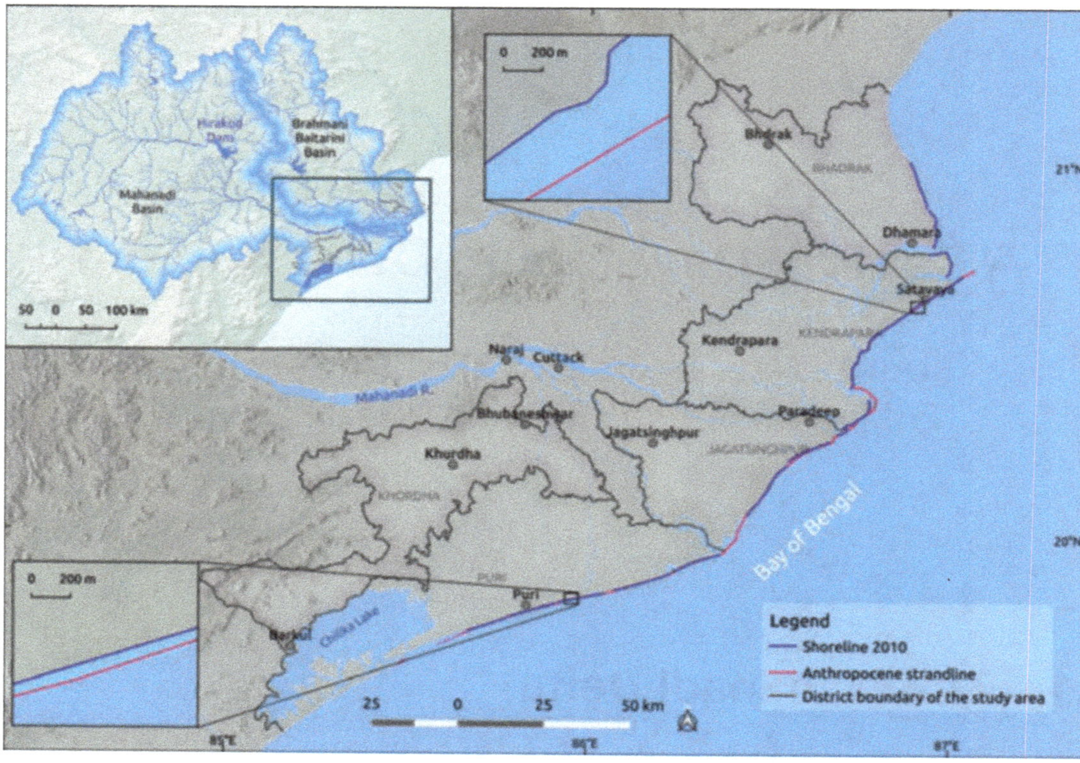

Fig. 3.1 Mahanadi Delta showing the extent of the catchment basin (inset) and the five coastal districts, with shoreline positions mapped in 1950 and 2010

covering 141,589 km^2 (WRIS 2011), 45% of which lies in the state of Odisha. The coastline of the delta is approximately 200 km long, extending from the Chilika lagoon in the south to the Dhamara River in the north. It has five coastal districts (Puri, Khordha, Jagatsinghpur, Kendrapara and Bhadrak) which constitute 83% of the delta area and have large areas below the five metre contour where floods due to cyclones and sea-level rise are expected to be important. These districts form the focus of the research presented here.

Indicators of the Anthropocene

During the Holocene period the delta shows evidence of substantial progradation and growth. The onset of the Anthropocene (post 1950) was indicated when this growth rate declined significantly, in line with a period of intense dam building (Somanna et al. 2016). This started with the construction of the multipurpose Hirakud Dam on the Mahanadi

River near Sambalpur in 1957, and has resulted in a total of 254 mainly small and medium scale dams within the drainage basin (WRIS 2014). The decline in sediment supply has been significant, amounting to 67% for the Mahanadi River and around 75% for the Bramhani River (Gupta et al. 2012). As a consequence, over 65% of the coastal margin is presently experiencing moderate to severe erosion, with increasing rates from south to north (Mukhopadhyay et al. 2018). For example, between 1990 and 2015, the Anthropocene shoreline (as mapped in 1950, see Fig. 3.1) receded at a rate often exceeding 10–15 m/year. Near the Mahanadi estuary, the rate has exceeded 50 m/year. Hence, erosion is now a key feature of the once prograding delta (Dandekar 2014; Mukhopadhyay et al. 2018).

Since the 1950s, the coastal districts of the delta have also witnessed rapid increase in population (especially Bhubaneswar city), growth of a port, industrial development along with increased groundwater extraction, small- and medium-scale irrigation projects and deforestation of mangroves. Based on the Census of India (2011a, b), the population in each coastal district has accelerated rapidly (see Fig. 3.2) reaching eight million in 2011 across the five coastal districts. The associated high population density is around 600 persons per km^2, growing at an estimated annual rate of 1.4% over the last two decades. Population projections anticipate this growth to continue until at least 2050 (Whitehead et al. 2015), although spatial variation in population distribution is noticeable (Fig. 3.2); Khordha is the most populated district with a population of 2.25 million (52% male and 48% female), whereas Jagatsinghpur district has the lowest population (1.13 million, 51% male and 49% female). The increase in population is allied with rapid urbanisation. The coastal city of Puri is famous for religious tourism and experiences an annual 10% rise in temporary population largely due to foreign tourists (Das 2013). On the outer fringe of Mahanadi Delta, the capital city of Bhubaneswar, established in 1948, registered the highest population growth rate in India during 1961–1971 with significant ongoing urban growth (Pathy and Panda 2012). Along with tourism, business and IT industries, education and health services make the area among the top three investment destinations in India (*The Economic Times* 2015). The port city of Paradip on the Mahanadi River, in

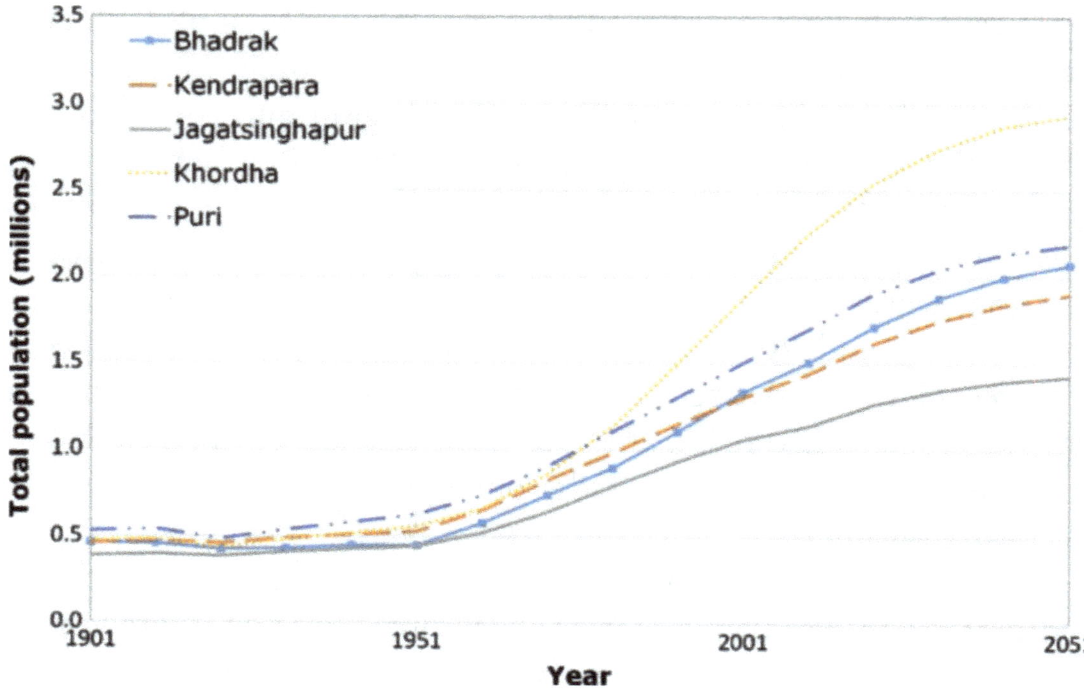

Fig. 3.2 Historical and projected decadal population growth for the five districts within the Mahanadi Delta (1901–2051) showing acceleration in the Anthropocene (post 1950s)

Jagatsinghpur district, is also emerging as a hub for major petroleum, chemicals and petrochemical investment regions.

As a consequence, the Mahanadi River system has experienced a substantial increase in the amount of untreated sewage and industrial effluent and rising levels of pollution (Radhakrishna 2001; Sundaray et al. 2006). During the peak monsoon discharge, high levels of organic pollution are indicated by Nitrogen and Phosphorus loading (Sundaray et al. 2006). Substantially low dissolved oxygen conditions have also been reported (Pradhan et al. 1998; Nayak et al. 2001) leading to more acidic conditions (Borges et al. 2003) with extremely low levels of pH being reported at various stations along the Mahanadi River in recent times (Panda et al. 2006; Sarma et al. 2012; Behera et al. 2014). The most strikingly low pH values have been observed in the Atharbanki creek water near the port at Paradip (pH = 3.2) with negligible seasonal variability (Sundaray et al. 2006). Such steady acidic input has a potential to affect agricultural productivity and contributes to the acidification of the Mahanadi Estuary with consequences for fisheries, aquaculture and tourism (Sarma et al. 2012; Bhattacharyya et al. 2015). Plastics, another key indicator of the Anthropocene period, are increasingly found in sediments and water bodies endangering fish, marine mammal and bird populations (e.g. in the Chilika lagoon [Sahu et al. 2014]).

More than 57% of the land area of the Mahanadi Delta is under cultivation. Agriculture currently (in 2011) occupies 68% and, with a subsidiary occupation of capture and culture fisheries, provides the subsistence for the predominantly rural population of the delta. The net area sown and gross copped area are 0.64 million hectares and 1.12 million hectares respectively (DAFP 2014). Agricultural intensity has been and is still being promoted by converting existing monocrop areas to grow two or three crops per year using irrigation water and chemical fertiliser input to boost crop production and yield (Ghosh et al. 2012; Srivastava et al. 2014; Pattanaik and Mohanty 2016).

The rapid urbanisation and development in agricultural practice over recent decades have left a mark on the environment. Land use analysis finds that between 2001 and 2011 alone, the delta has lost 4600 hectares of agriculture land and 360 hectares of pristine mangrove forest to urban/rural settlements and aquaculture. A probabilistic land cover change assessment for 2030 (Fig. 3.3) under a Business as Usual (BAU) scenario (see Kebede et al. 2018) suggests that the double and triple crop areas in the delta are likely to increase by 12 and 21%, respectively, with substantial (48% when compared to 2011) reduction in mangrove areas (and associated unique ecosystem services). This change in agriculture is reflected in the economics of the delta with the agriculture and animal husbandry sector contributing only 17% (2012–2013) to the Gross State Domestic Product (GSDP) at current prices and the share of the agricultural sector in the states gross domestic product declining over the last decade (P & CD 2014, 2015). Lower profitability of agriculture coupled with reducing yield per hectare is considered the major reason for this decline. Climate change and associated hazards have the potential to further reduce agricultural yield by 2050, affecting food security and reducing the GDP of the delta population by around 0.27% (Cazcarro et al. 2018) (see Chapter 8).

Fisheries, both marine capture and culture, are another important livelihood option for the people in the delta. Before the 1950s,

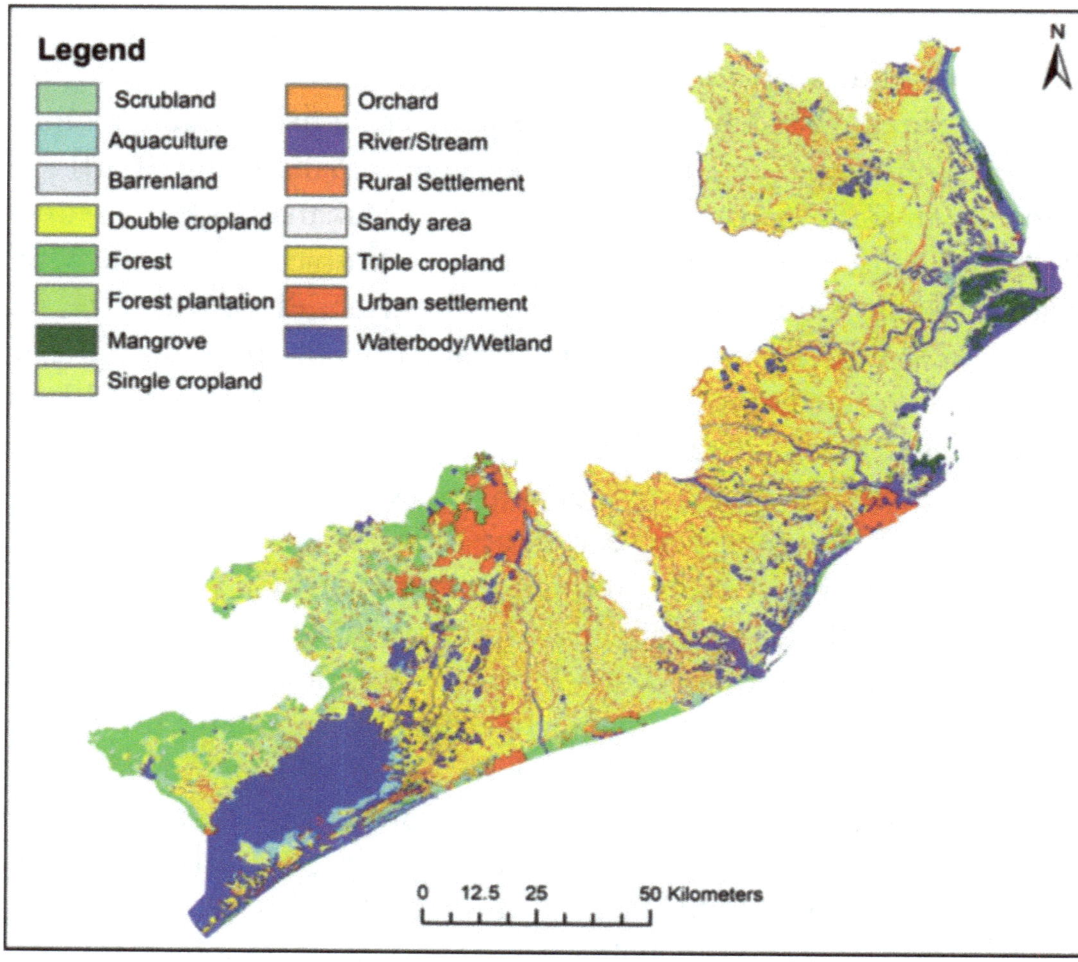

Fig. 3.3 Projection of land cover change in the Mahanadi Delta for 2030 under a Business as Usual scenario

people used to depend on subsistence fishing in the river and the estuaries using boats to supplement their daily protein requirements. In the capture fisheries sector, motorised boats were introduced in the late 1950s but only became popular with the increasing availability of outboard engines during the 1980s. This is reflected in the increase in average annual marine catch from 5000 tonnes (1950s) to 120,000 tonnes (1990s). However, in spite of the introduction of technology in the form of improved engines and fishing gears, the marine catch still shows abrupt fluctuation due to variability of the monsoon, river discharge and other oceanographic parameters. Over the last two decades, while the catch of prawns (*Penaeid* and non *Penaeid* species) or Ribbon fish showed a steady increase, the prize catch of Hilsa, Silver Pomfret, Bombay Duck or Indian Spanish Mackerel showed a wide variation. For the future, the impact of climate change on marine fisheries of northern Bay of Bengal analysed using integrated modelling (Fernandes et al. 2016a, b; Cazcarro et al. 2018; Lauria et al. 2018) is expected to reduce marine fish production in the Bay of Bengal region between 3 and 9% by 2100 under a BAU scenario. The socio-economic implications of this decline indicate that, if combined with unsustainable fishing practices, there would be significant loss in the GDP of the Mahanadi Delta region by 2050 (Lauria et al. 2018).

Extreme Events and Climate Change

Climatic extremes have a potential to affect the delta adversely. Odisha is the fifth most flood-prone state in India with the delta exposed to frequent floods and waterlogging. In addition to heavy rainfall, cyclonic winds and tidal flows also cause flooding in coastal areas with flooding usually lasting for five to fifteen days in the coastal districts. While low to moderate intensity floods are often treated as a boon by the coastal community due to the arrival of high fertility soils, it is the high-intensity floods that adversely affect the lives, livelihood and food security of the coastal community (F & ED 2010, 2018).

The Mahanadi Delta is situated in the most cyclone-prone region of India. Historical data on cyclones in Odisha indicate high disaster losses due to cyclone and surges in the Anthropocene period (Chittibabu et al. 2004) with eight high-intensity flooding events reported during the period 2001–2015 (Ghosh et al. 2019). One of the most extreme events experienced was the 1999 Odisha Super cyclone Kalinga which had an estimated maximum wind speed of 260–270 km/hr generating a surge of more than 6 metres (20 feet) which travelled 20 km inland (Kalsi 2006). This along with heavy rainfall led to substantial loss of life and damage to property. In the five deltaic districts, 9078 lives were lost, 445,595 houses collapsed, 13,762 houses were washed away and around 0.7 million hectares of agriculture land was affected (PCD 2004). In recent times, thanks to the improved cyclone and flood warning, evacuation and disaster management procedures and community preparedness, loss of life and property can be minimised (Padhy et al. 2015) as evidenced during the very severe cyclonic storms Phailin (2013) and HudHud (2014). Simulations from a regional climate model suggest that the frequency and intensity of severe cyclones are likely to increase along with extreme sea levels in the later part of this century when compared to a baseline scenario (1961–1990) (Unnikrishnan et al. 2011).

Future climate data from regional modelling (Macadam and Janes 2017) also indicates that precipitation along with high rainfall events may increase significantly in the later part of this century. Using daily discharge simulations from bias corrected CNRM-C5 data between 2021 and 2099, it is observed that the number of high discharge events (those exceeding 20,000 cumec, capable of generating flood in the delta) is likely to increase under a BAU scenario. The greatest impact of these high discharge events and flooding would be on agricultural land along with a number of urban areas across the delta. Table 3.1 shows the likely extent of flooding and coastal inundation of agricultural land in the delta by the end of the century under this scenario.

In addition, the rate of relative sea-level rise in the last decade has increased to nearly 6 mm/year (Mukhopadhyay, personal communication computed from PSMSL data [http://www.psmsl.org/]). Forecasting the future coastline for the years 2020, 2035 and 2050, Mukhopadhyay

Table 3.1 Spatial extent of agricultural area inundation in the Mahanadi Delta by 2100 under a Business as Usual scenario

Blocks/sub-districts	Area of cropland (km^2)	
	Projected to be flooded in 100 year fluvial floods	Projected to be flooded coastal storm surges
Chandabali	418	421
Basudebpur	298	256
Mahakalpara	290	280
Rajnagar	251	263
Tihidi	212	176
Ersama	211	148
Pattamundai	206	152
Rajkanika	–	183

et al. (2018) estimate that about 26 coastal *mouzas* (small administrative blocks) out of 87 (approximately 30%) with average population density greater than 400 per square kilometre would be susceptible to acute erosion by the year 2050 and this may lead to displacement of people from the coast.

3.2 Key Issues for the Mahanadi Delta

Migration

Throughout history, deltas have been a preferred destination for people to migrate and settle thanks to the abundance of natural resources and livelihood opportunities (Szabo et al. 2016). However, the Mahanadi Delta itself has seen a net trend of out-migration. In the state of Odisha, migration (mostly internal) can be permanent, seasonal or circular, and, as it is an ongoing process, it can be difficult to identify causes.

From a study of perceptions of environmental stress of migrant households, it emerges that hazard events such as flooding and droughts act as 'stressors' and motivate individuals/households to consider migration as an option. This is supported by analysis and mapping of risk and net-migration at sub-district level (Fig. 3.4) which includes the common environmental stressors of flooding, cyclone and coastal erosion. Several coastal sub-districts are shown to be adversely affected by climatic hazards, have a low level of economic growth, exhibit high risk and, critically, experience greater out-migration. The sub-districts of Dhamnagar, Ersama, Balikuda and Tihidi are biophysically and socio-economically at very high risk and out-migration dominates (Fig. 3.4) whereas, several sub-districts at comparatively lower risk (e.g. Khordha, Puri) and urban growth centres mostly act as net receiving areas of migrants. Khordha, which is the most urbanised district in Odisha (43% urban population), Puri (famous destination for religious tourism) and Paradip (a growing seaport) emerge as preferred destinations of migrants as they offer economic opportunities for migrants from adjoining rural communities (Das et al. 2016).

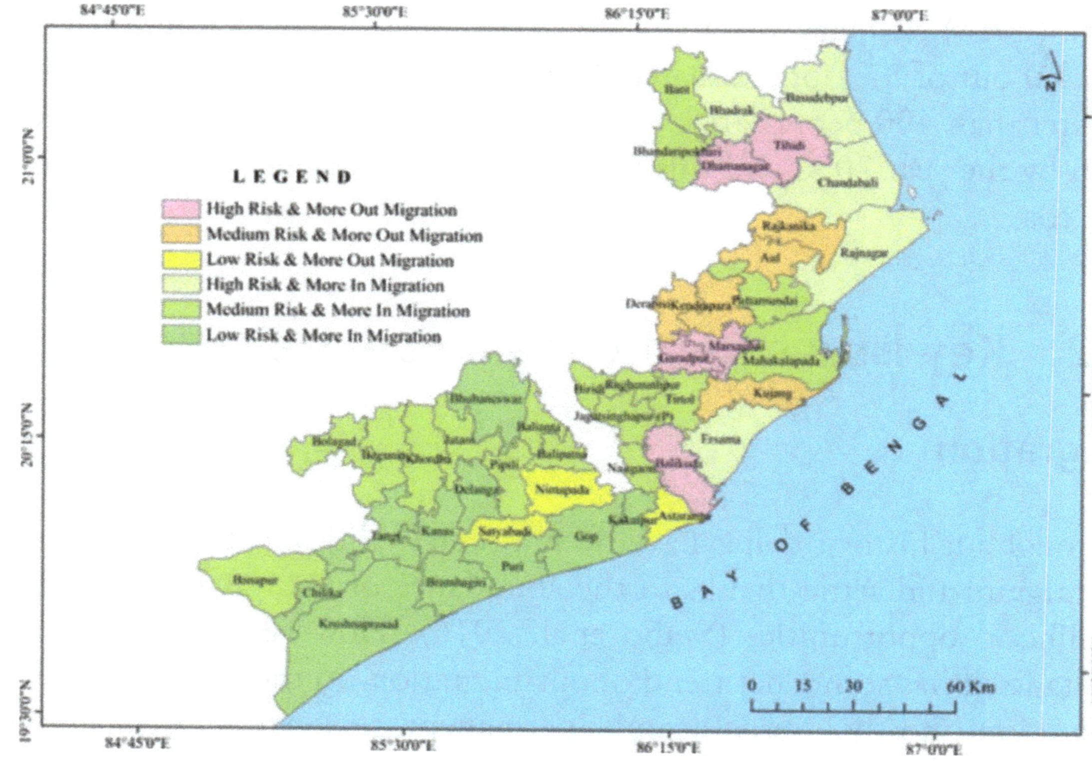

Fig. 3.4 Multi-hazard risk and migration map for the Mahanadi Delta

However, in a household survey carried out in 2016 (see Table 3.2), over 40% of respondents reported that the main reason behind migration is economic, with the majority of migrants moving in search of better employment opportunities. The second most frequently mentioned reason (nearly 20%) is associated with education; to pursue a degree or obtain training. The survey also shows that the dominant nature of migration is seasonal (62%). People migrate to major cities and different states of India once or twice a year depending on the season. Migrants are mostly from agricultural households where monthly income is low and where household size is large (more than 6 members). It is possible that low returns from the existing livelihoods of agriculture and fisheries are triggering migration in the hope to generate alternative livelihoods.

The propensity to migrate is highest in the 21–30 age group and people with secondary and tertiary education generally tend to migrate more than those with lower levels of education. It is most common for single male members of a family to migrate; women either migrate with family members, or remain as female heads of households. Male-headed families

Table 3.2 Patterns of migration in the Mahanadi Delta based on a household survey carried out in 2016

Characteristics		Responses (%)
Type	Seasonal migration	62.1
Frequency	1–2 times	54.2
Duration	3–6 months	39.7
Scale	Internal migration	99.2
Destination		
State	Odisha, West Bengal, Karnataka, Tamil Nadu	69.7
District	Khordha, Kolkata, Puri	60.4
City	Bhubaneswar, Puri (M), Khordha (M)	75.3
Current migrant's characteristics		
Male migrants		
Age	21–40	60.6
Marital status	Never married	52.9
Education	Secondary	49.7
Livelihood	Factory worker, regular salaried employee, construction worker	78.1
Monthly income	Rs. 10,000 and below	71.6
Female migrants		
Age	21–40	50.0
Marital status	Currently married	54.8
Education	Secondary	43.9
Livelihood	Unpaid home carer, student	88.2
Monthly income	No income	91.8
Reasons		
First	Seeking employment	43.6
Second	Seeking education	19.4
Third	Family obligations/problems	13.5
Remittances		
Type	Money	42.0
Frequency	Monthly	47.4
Amount	Rs. 5000 and below	70.3
Uses of remittances	Daily consumption (food, bills)	71.0

dominate in the delta (1225 [87%] male as opposed to 189 [13%] female), but the increase in the number of female-headed household is emerging as a feature of Anthropocene in the Mahanadi Delta; it is estimated from the household surveys that the number of migrant-sending households will increase in the delta to 38% in the near future resulting in more married women 'left behind' as female heads of households.

These female-headed households often have a more difficult time during the extreme events than male heads due to family care responsibilities, lower incomes, lower resilience or adaptive capacity. A key conclusion from the survey is that female-headed households experienced more monetary losses due to failure of crop, livestock and equipment damages as well as loss of life, during the extreme events than their male-headed counterparts. More than 37% of the female heads have no income and 47% have income less than INR 3000 per month. Sixty per cent of female heads are found to be widowed. In fact, widows, dominantly of mature age (54% are over 60 years old) and with no education (>50%) are often dependent on pension schemes (widow pension, old age pension) provided by the government or are supported by relatives. This suggests that female-headed households living in physically most vulnerable conditions in the delta are socio-economically more vulnerable than the male-headed households.

Monthly remittances sent by migrant family members can alleviate the vulnerable status of households, at least marginally. Remittances enable recipients to pay for daily consumption (food, bills), education and health, and to maintain or improve their standard of living. Households with a migrant member are economically better placed than those without migrants. This is more prominent in the case of female-headed households. Monthly Per Capita Expenditure (MPCE) of female-headed households with migrant members are found to be higher (Rs. 2355) than those without any migrant member (Rs. 1473). More than 60% of total respondents felt migration is beneficial as this improves the socio-economic status of migrants and migrant-sending households. The exchange of money, knowledge and ideas between migrant's place origin and destinations offers further opportunities for reducing the socio-economic and biophysical vulnerabilities for communities within the delta.

Adaptation

Adaptation activities, both autonomous and planned are taking place in the delta in relation to vulnerability reduction, disaster risk reduction and building social-ecological resilience to natural hazards and environmental changes (see DECCMA 2018). A notable case of planned adaptation due to environmental change is the Satabhaya Gram *panchayat* of the Kendrapara district in the delta. After a loss of 65% of land area due to erosion, 571 families were identified by the Government of Odisha to be resettled and rehabilitated in Bagpatiya village in 2010 (R & DM Department 2011). However, while more than 50% of families have been resettled, full implementation of the plan is yet to be achieved.

Survey results, at the household level, recorded adaptation activities can be broadly subdivided into three categories: disaster risk reduction, livelihood assistance and infrastructure building. The first category includes capacity building and training in various forms to improve resilience/adaptive capacity of the individual or community for reduction of disaster risk, which is observed to be a dominant mode of adaptation in the Mahanadi Delta (Tompkins et al. 2017); the second category of livelihood related activities include climate tolerant crops, mixed farming, irrigation and water resources augmentation and fishing new breeds; the third category of infrastructure related adaptation involves construction of embankments, house relocation or upgrade and cyclone shelters. The impact of these adaptations is reflected in improved agricultural productivity, food security and efficient management of water resources and enhanced income (Hazra et al. 2016). Use of these techniques is spatially variable across the delta (see Fig. 3.5).

Capacity building and training, climate-tolerant crops, assistance from government and NGOs and structural protection measures like embankment or cyclone shelters were seen to be effective, with loans, cutting down trees and the use of mixed farming methods deemed less successful by those surveyed. For those working at State and District level, the topmost criterion for successful adaptation is the improvement of the capacity of the local institutions to manage environmental disasters and changes. Odisha is exposed to floods and cyclones, and this indicates the need for enhancing capacity to cope with these disasters at all levels across the delta.

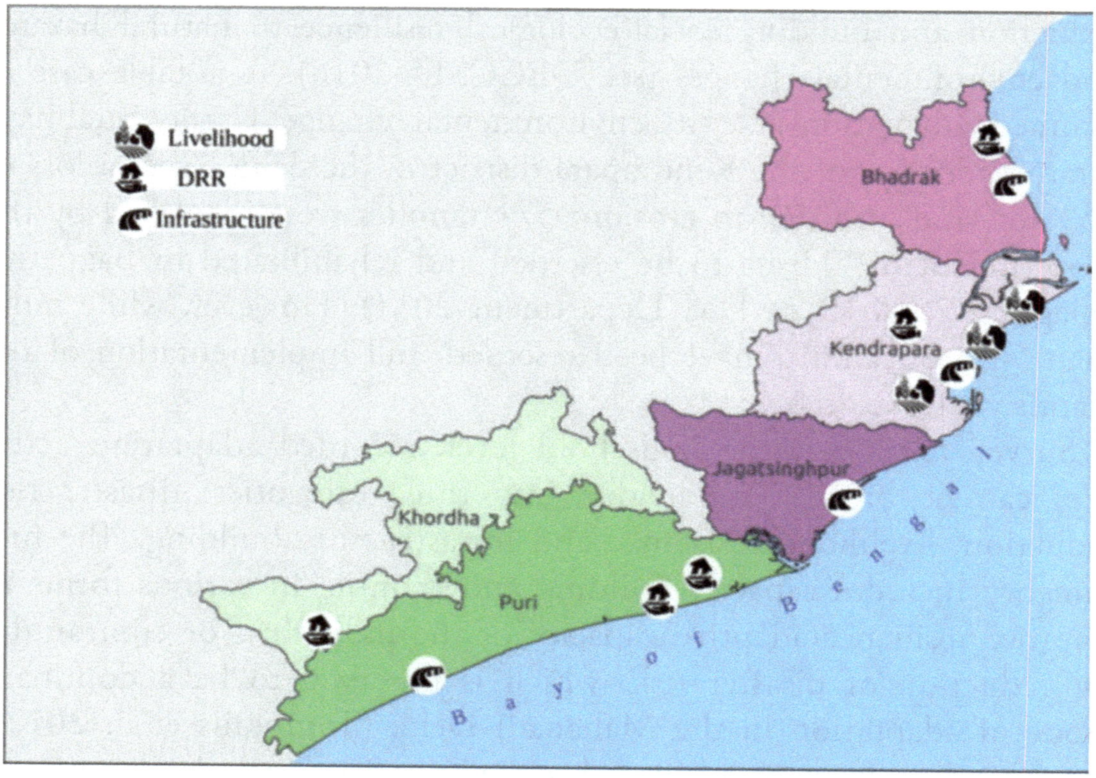

Fig. 3.5 Spatial variability of adaptation activities in the Mahanadi Delta

During the household surveys, the opinion of both male and female respondents were collected on 21 adaptation options that they may/may not practice, the scale of relative success and the preference or intention of the respondents for a particular type (Fig. 3.6 upper). Other than infrastructure and agriculture/fishing (livelihood) related activities, the soft options for disaster reduction related activities such as training, capacity building, assistance, loan, etc. are included in Institutional Support. Migration, returned migration or women working outside the village have been considered under mobility. The responses are also plotted against the percentage of success (abscissa) and percentage preference (ordinate) with the size of the bubble varying with the frequency of practice (Fig. 3.6 lower). From these results it can be observed that there are gender differences in the practiced adaptation activities and often the most practiced adaptation activities are not always the most successful one.

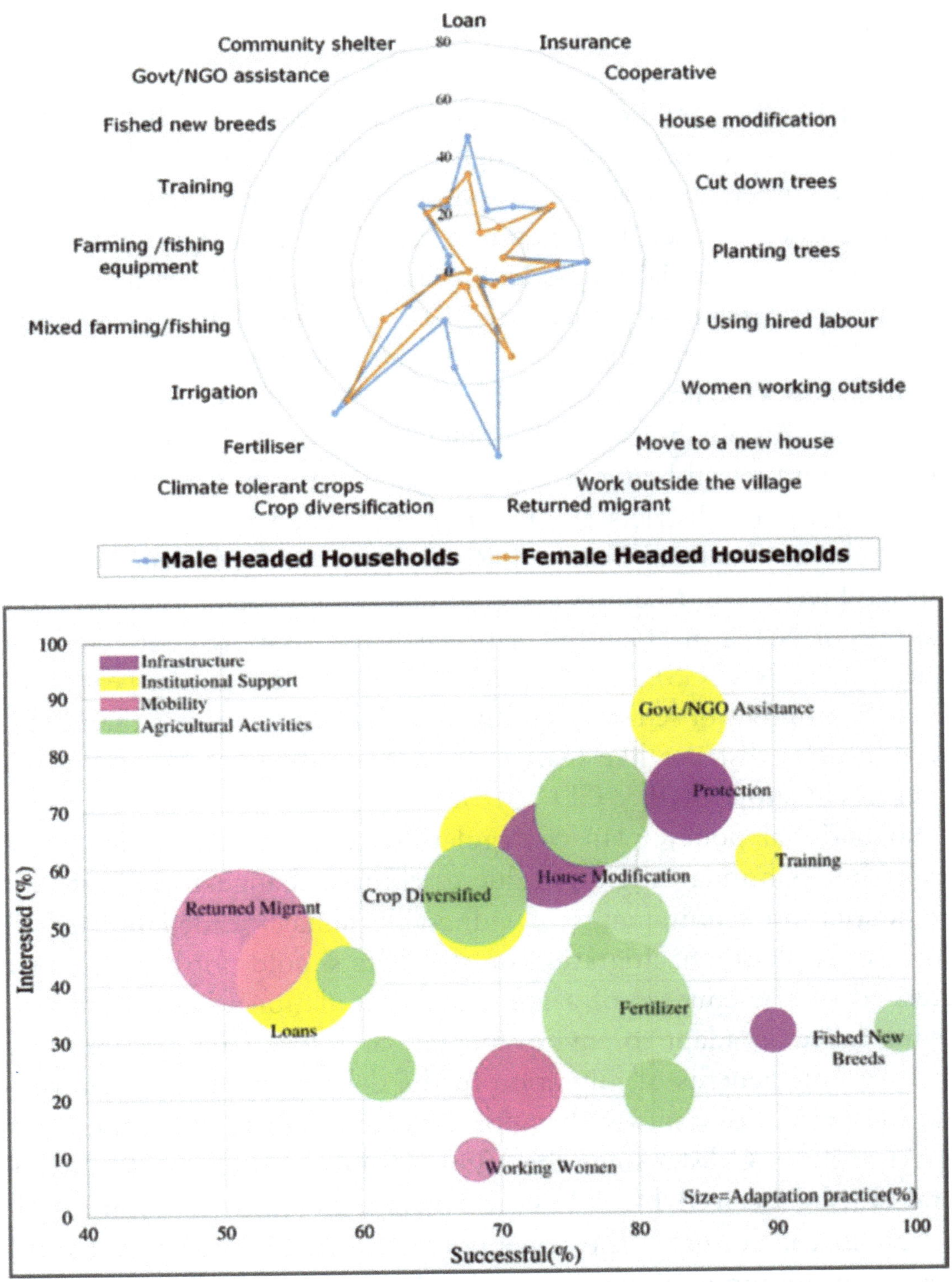

Fig. 3.6 Gender differentiated adaptation activities (upper) and adaptation practice, success and preference (lower)

Policy

In Odisha, there are no specific policies focussed on the Mahanadi Delta. This part of the State, though both biophysically and socio-economically vulnerable to climatic and demographic changes, have never been addressed as a separate planning unit. The most significant coastal project with a potential long-standing impact on the sustainable development of the delta is the Integrated Coastal Zone Management Project (ICZMP), being implemented in two coastal stretches—Paradeep to Dhamra and Gopalpur to Chilika. However, the investigation of the biophysical sustainability of the delta as a physiographic entity and socio-economic well-being of the deltaic community was not included. The delta coastal districts are also not considered in livelihood generation schemes such as National Rural Livelihood Mission (NRLM)/ Odisha Livelihood Mission (OLM) and employment guarantee programme of Mahatma Gandhi National Rural Employment Guarantee Scheme (MGNREGS). The NRLM also reports that in the year 2013–2014 the number of self-help groups in delta districts is less than half of the non-delta districts like Ganjam and Mayurbhanj functioning with government support (P & CD 2015).

No national policy, plan or regulations covering adaptation activities addresses human migration due to climate change and disasters, or the adaptation requirements of vulnerable female-headed households. However, a picture of migration on the delta should not be solely characterised by the concept of a vulnerable population forced to migrate, having failed to adapt to environmental variability. This research identified various benefits of migration in the delta, with remittances being one of them. Data shows that domestic remittances have risen most significantly in Odisha since the 1990s (Tumbe 2011). In 2007–2008, rural Odisha received 14.25 billion dollars as domestic remittances, 6th highest in the country. The benefits of remittances positively contributed to autonomous climate adaptation by the community.

Another inadequacy of present policies of climate change adaptation and disaster risk reduction is the limited consideration of gender in policies and more generally a lack of a gendered database for an effective yet differentiated planning. Such differentiated policies would be necessary to address the rising number of female-headed households in the delta. Making certain of gender equity in decision-making is very important in migration contexts, especially when the out-migrants are generally adult male members of the family. Empowerment of women through various skill building and other trainings will help in developing the adaptive capacity of this vulnerable group.

3.3 The Future for the Mahanadi Delta

In the Anthropocene, where human beings have the capacity to alter the pathway of natural changes, the future is what humans make it. Future projections under a BAU scenario are not encouraging for the Mahanadi Delta. Higher sea surface temperatures (a rise of potentially 2.3–2.9 °C [Fernandes et al. 2016b]) and sea levels along with an increasing number of high rainfall events, particularly in the later part of the century, has a potential for increased number of flood events endangering life and livelihoods of the delta community. Climate change and climatic shocks also have the potential to reduce crop yield in the delta. Based on the IPCC's Shared Socio-economic Pathway 2 scenario for the Mahanadi Delta, Arto et al. (2019) estimated, that, by 2050, the economic loss from agriculture could be about 5% of GDP per capita. However if loss of infrastructure is considered, climate change and climatic shocks (flood, cyclone, etc.) may lead to cumulative per cent loss in GDP per capita of about 11% in the delta (Arto et al. 2019). At the same time, with the projected loss of fisheries production (Fernandes et al. 2016a), the socio-economic integrated model indicates that losses from the fishery sector alone to be around 0.25% of the total GDP in the delta by 2050 (Lauria et al. 2018). These together impose a serious constraint on the livelihood of the delta community in future.

Like most other deltas in the world, Mahanadi Delta, is under threat from sea-level rise and sediment starvation (Chapter 5). The potential consequences of extreme events are illustrated by surge-inundation from the 1999 Super cyclone when the district of Jagatsinghpur suffered over eight thousand fatalities (see Table 7.6 in PCD 2004). As indicated earlier, 65% of the deltaic coast is also currently experiencing varying degree of erosion, a situation which is expected to worsen by 2050 (Mukhopadhyay et al. 2018), implying compounded threat to the coastal habitations particularly between Puri to Paradip and promoting increased out-migration of people from the coast.

Whether the Mahanadi Delta will be able to withstand adverse environmental hazards in the Anthropocene and develop sustainably will largely depend on global climate change mitigation action (see Brown et al. 2018) alongside regional, national and local adaptation planning. As the latter are currently limited in scope and application, an immediate challenge is to generate management plans which aim to develop holistic pathways to sustainable management of biophysical and human resources of the delta in near future.

3.4 Discussion and Conclusion

This research looks at the changes in the Mahanadi catchment basin since the onset of the Anthropocene (i.e., since 1950), focusing on the evolution of the Mahanadi Delta, both in terms of biophysical and socio-ecological change over time including delta management and policy evolution. The once prograding delta of the Holocene is now retreating due to sediment starvation and sea-level rise, and experiencing accelerated population growth, decline in income from agriculture or fisheries, increasing pollution in the river system with potential of acidification of estuaries, proliferation of plastics in the environment, degradation of mangroves with loss of biodiversity and human migration. Thus the Mahanadi Delta is now shrinking and a more challenging situation is emerging. The most vulnerable communities often suffer disproportionately at times of natural disasters with potential loss of life, livelihoods and assets. The people living in the Mahanadi Delta are forced to cope with frequent disasters, but recent progress in warnings, evacuation and shelters seems to have reduced losses. Urban areas in the delta are expanding and there is rural to urban migration which can be expected to continue. These urban areas will have important implications for the future of the delta.

For policymakers at both the State and District level, the topmost criterion for successful adaptation is the improvement of the capacity of the local institutions to manage environmental disasters and changes. This research indicates a positive correlation between perceived success of adaptation activity and people's intention to undertake them. However, the most practiced adaptation activities are not always the most successful in the Mahanadi Delta. Adaptation activities are not always gender-sensitive with the effect that men are benefiting more than women. Thus it has to be ensured that the most vulnerable groups including the elderly, the severely poor, the physically challenged and women are involved in the development of migration-related adaptation as stakeholders. All stakeholders need to be involved for proper planning and implementation of adaptation plans and for accrual of benefits from there.

This research has the following three key observations for the Mahanadi Delta: (i) assessing the Mahanadi Delta region as a unit of planning and implementation offers opportunities to enable coherent policy responses to reduce the risk of climate change to populations by supporting gender-sensitive adaptations; (ii) a variety of types of adaptations should be used to reduce climate change risk; and (iii) to encourage migration as an adaptation, targeting trapped populations with skill development (as observed in Pattamundai, Kendrapara district) can improve their migration capability. Alternatively, designing income-generating opportunities like skills training, livelihood programmes and development initiatives for those who remain behind in areas affected by disasters/climate change will reduce their vulnerability and increase their ability to cope during an extreme weather event.

In the Anthropocene, the major challenge for the rural areas of the Mahanadi Delta is to restore or maintain the natural delta dynamics as far as possible, to combat the cumulative threat of sea-level rise and land subsidence and address the water-sediment-pollution-biodiversity interactions, while maintaining the socio-economy of the delta in an integrated and inclusive way to make the delta sustainable for the future generations. The interactions of the rural delta areas with the urban delta areas also need to be considered.

References

Arto, I., García-Muros, X., Cazcarro, I., González, M., Markandya, A., & Hazra, S. (2019). The socioeconomic future of deltas in a changing environment. *Science of the Total Environment, 648,* 1284–1296. https://doi.org/10.1016/j.scitotenv.2018.08.139.

Behera, B. C., Mishra, R. R., Patra, J. K., Dutta, S. K., & Thatoi, H. N. (2014). Physico chemical properties of water sample collected from mangrove ecosystem of Mahanadi River Delta, Odisha, India. *American Journal of Marine Science, 2*(1), 19–24. https://doi.org/10.12691/marine-2-1-3.

Bhattacharyya, R., Ghosh, N. B., Mishra, K. P., Mandal, B., Rao, S. C., Sarkar, D., et al. (2015). Soil degradation in India: Challenges and potential solutions. *Sustainability, 7*(4), 3528–3570. https://doi.org/10.3390/su7043528.

Borges, A. V., Djenidi, S., Lacroix, G., Théate, J., Delille, B., & Frankignoulle, M. (2003). Atmospheric CO2 flux from mangrove surrounding waters. *Geophysical Research Letters, 30*(11), 1558. https://doi.org/10.1029/2003GL017143.

Brown, S., Nicholls, R. J., Lázár, A. N., Hornby, D. D., Hill, C., Hazra, S., et al. (2018). What are the implications of sea-level rise for a 1.5, 2 and 3 °C rise in global mean temperatures in the Ganges-Brahmaputra-Meghna and other vulnerable deltas? *Regional Environmental Change, 18*(6), 1829–1842. https://doi.org/10.1007/s10113-018-1311-0.

Cazcarro, I., Arto, I., Hazra, S., Bhattacharya, R. N., Adjei, P. O.-W., Ofori-Danson, et al. (2018). Biophysical and socioeconomic state and links of deltaic areas vulnerable to climate change: Volta (Ghana), Mahanadi (India) and Ganges-Brahmaputra-Meghna (India and Bangladesh). *Sustainability, 10*(3), 893. https://doi.org/10.3390/su10030893.

Census of India. (2011a). *Primary Census Abstract.* Office of the Registrar General and Census Commissioner, Government of India. http://censusindia.gov.in/pca/default.aspx. Last accessed 5 November 2018.

Census of India. (2011b). *Table A2 area, population, decennial growth rate and density for 2001 and 2011 at a glance for West Bengal and the Districts*. Office of the Registrar General and Census Commissioner, Government of India. http://censusindia.gov.in/2011-prov-results/prov_data_products_wb.html. Last accessed 5 November 2018.

Chittibabu, P., Dube, S. K., Macnabb, J. B., Murty, T. S., Rao, A. D., Mohanty, U. C., et al. (2004). Mitigation of flooding and cyclone hazard in Orissa, India. *Natural Hazards, 31*(2), 455–485. https://doi.org/10.1023/B:NHAZ.0000023362.26409.22.

DAFP. (2014). *Odisha Agriculture Statistics 2012–13*. Bhubaneswar, Odisha, India: Directorate of Agriculture and Food Production (DAFP), Government of Odisha.

Dandekar, P. (2014). *Shrinking and sinking deltas: Major role of dams in delta subsidence and effective sea level rise*. Delhi, India: South Asia Network on Dams Rivers and People (SANDRP). http://www.indiaenvironmentportal.org.in/files/file/Shrinking_and_sinking_delta_major_role_of_Dams_May_2014.pdf. Last accessed 30 September 2018.

Das, S. K. (2013, August/September). Growth and prospects of Odisha tourism: An empirical study. *Odisha Review*, 125–134. New Delhi, India: Government of India. http://magazines.odisha.gov.in/Orissareview/2013/aug-sept/engpdf/126-135.pdf. Last accessed 10 October 2018.

Das, S., Hazra, S., Ghosh, T., Hazra, S., & Ghosh, A. (2016, May 10–13). *Migration as an adaptation to climate change in Mahanadi Delta*. Adaptation Futures 2016. Rotterdam, The Netherlands. https://edepot.wur.nl/381486. Last accessed 10 October 2018.

DECCMA. (2018). *Climate change, migration and adaptation in deltas: Key findings from the DECCMA project* (Deltas, Vulnerability and Climate Change: Migration and Adaptation [DECCMA] Report). Southampton, UK: DECCMA Consortium. https://www.preventionweb.net/publications/view/61576. Last accessed 27 November 2018.

F & ED. (2010). *Odisha Climate Change Action Plan 2010–2015*. Bhubaneswar, Odisha, India: Forestry and Environment Department (F & ED), Government of Odisha. http://climatechangecellodisha.org/publication.html. Last accessed 7 November 2018.

F & ED. (2018). *Odisha Climate Change Action Plan Phase II (2018–23)*. Bhubaneswar, Odisha, India: Forestry and Environment Department (F & ED), Government of Odisha. http://climatechangecellodisha.org/publication.html. Last accessed 30 November 2018.

Fernandes, J. A., Kay, S., Hossain, M. A. R., Ahmed, M., Cheung, W. W. L., Lazar, A. N., et al. (2016a). Projecting marine fish production and catch potential in Bangladesh in the 21st century under long-term environmental change and management scenarios. *ICES Journal of Marine Science, 73*(5), 1357–1369. https://doi.org/10.1093/icesjms/fsv217.

Fernandes, J. A., Papathanasopoulou, E., Hattam, C., Queirós, A. M., Cheung, W. W. W. L., Yool, A., et al. (2016b). Estimating the ecological, economic and social impacts of ocean acidification and warming on UK fisheries. *Fish and Fisheries, 18*(3), 389–411. https://doi.org/10.1111/faf.12183.

Ghosh, A., Das, S., Ghosh, T., & Hazra, S. (2019). Risk of extreme events in delta environments: A case study of the Mahanadi Delta. *Science of the Total Environment, 664*, 713–723.

Ghosh, S., Verma, H. C., Panda, D. K., Nanda, P., & Kumar, A. G. (2012). Irrigation, agriculture, livelihood and poverty linkages in Odisha. *Agricultural Economics Research Review, 25*(1), 99–105.

Gupta, H., Kao, S.-J., & Dai, M. (2012). The role of mega dams in reducing sediment fluxes: A case study of large Asian rivers. *Journal of Hydrology, 464–465*, 447–458. https://doi.org/10.1016/j.jhydrol.2012.07.038.

Hazra, S., Dey, S., & Ghosh, A. K. (2016). *Review of Odisha State adaptation policies, Mahanadi Delta* (Deltas, Vulnerability and Climate Change: Migration and Adaptation [DECCMA] Working Paper). Southampton, UK: DECCMA Consortium. https://generic.wordpress.soton.ac.uk/deccma/resources/working-papers/. Last accessed 6 August 2018.

Kalsi, S. R. (2006). Orissa super cyclone: A synopsis. *Mausam, 57*(1), 1–20.

Kebede, A. S., Nicholls, R. J., Allan, A., Arto, I., Cazcarro, I., Fernandes, J. A., et al. (2018). Applying the global RCP–SSP–SPA scenario framework at sub-national scale: A multi-scale and participatory scenario approach. *Science of the Total Environment, 635*, 659–672. https://doi.org/10.1016/j.scitotenv.2018.03.368.

Kumar, K. V., & Bhattacharya, A. (2003). Geological evolution of Mahanadi Delta, Orissa using high resolution satellite data. *Current Science, 85*(10), 1410–1412.

Lauria, V., Das, I., Hazra, S., Cazcarro, I., Arto, I., Kay, S., et al. (2018). Importance of fisheries for food security across three climate change vulnerable deltas. *Science of the Total Environment, 640–641*, 1566–1577. https://doi.org/10.1016/j.scitotenv.2018.06.011.

Macadam, I., & Janes, T. (2017). *Validation of Regional Climate Model simulations for the DECCMA project* (Deltas, Vulnerability and Climate Change: Migration and Adaptation [DECCMA] Working Paper). Southampton, UK: DECCMA Consortium. https://generic.wordpress.soton.ac.uk/deccma/resources/working-papers/. Last accessed 10 October 2018.

Mukhopadhyay, A., Ghosh, P., Chanda, A., Ghosh, A., Ghosh, S., Das, S., et al. (2018). Threats to coastal communities of Mahanadi Delta due to imminent consequences of erosion—Present and near future. *Science of the Total Environment, 637–638,* 717–729. https://doi.org/10.1016/j.scitotenv.2018.05.076.

Nayak, B. B., Das, J., Panda, U. C., & Acharya, B. C. (2001). Industrial effluents and municipal sewage contamination of Mahanadi estuarine water, Orissa. In *Proceedings* (pp. 77–96). New Delhi, India: Allied Publishers.

Padhy, G., Padhy, R. N., Das, S., & Mishra, A. (2015). A review on management of cyclone Phailin: Early warning and timely action saved lives. *Indian Journal of Forensic and Community Medicine, 2*(1), 56–63.

Panda, U. C., Sundaray, S. K., Rath, P., Nayak, B. B., & Bhatta, D. (2006). Application of factor and cluster analysis for characterization of river and estuarine water systems—A case study: Mahanadi River (India). *Journal of Hydrology, 331*(3), 434–445. https://doi.org/10.1016/j.jhydrol.2006.05.029.

Pathy, A. C., & Panda, G. K. (2012). Modeling urban growth in Indian situation—A case study of Bhubaneswar City. *International Journal of Scientific & Engineering Research, 3*(6), 1–7.

Pattanaik, F., & Mohanty, S. (2016). Growth performance of major crop groups in Odisha agriculture: A spatiotemporal analysis. *Agricultural Economics Research Review, 29*(2), 225–237.

PCD. (2004). *Human development report.* Prepared by Nabakrushna Choudhury Centre for Development Studies. Bhubaneswar, Odisha, India: Planning and Coordination Department (PCD) Government of Odisha. http://www.planningcommission.nic.in/plans/stateplan/sdr_pdf/shdr_ori04.pdf. Last accessed 10 October 2018.

P & CD. (2014). *Economic Survey Report 2013–2014.* Bhubaneswar, Odisha, India: Planning and Convergence Department (P & CD), Government of Odisha. http://pc.odisha.gov.in/. Last accessed 7 November 2018.

P & CD. (2015). *Economic Survey Report 2014–2015.* Bhubaneswar, Odisha, India: Planning and Convergence Department (P & CD), Government of Odisha. http://pc.odisha.gov.in/. Last accessed 7 November 2018.

Pradhan, S. K., Patnaik, D., & Rout, S. P. (1998). Ground water quality—An assessment around a phosphatic fertiliser plant at Paradip. *Indian Journal of Environmental Protection, 18*(10), 769–772.

Radhakrishna, I. (2001). Saline fresh water interface structure in Mahanadi Delta region, Orissa, India. *Environmental Geology, 40*(3), 369–380. https://doi.org/10.1007/s002540000182.

R & DM Department. (2011). *Package for relocating the villagers of Satabhaya and Kanhupur in Bagapatia.* Letter R&REH-45/11-18573/R&DM. Bhubaneswar,

Odisha: Revenue and Disaster Management Department, Government of Odhisa. http://revenueodisha.gov.in. Last accessed 5 November 2018.

Sahu, B. K., Pati, P., & Panigrahy, R. C. (2014). Environmental conditions of Chilika Lake during pre and post hydrological intervention: An overview. *Journal of Coastal Conservation, 18*(3), 285–297. https://doi.org/10.1007/s11852-014-0318-z.

Sarma, V. V. S. S., Krishna, M. S., Rao, V. D., Viswanadham, R., Kumar, N. A., Kumari, T. R., et al. (2012). Sources and sinks of CO2 in the west coast of Bay of Bengal. *Tellus B: Chemical and Physical Meteorology, 64*(1). https://doi.org/10.3402/tellusb.v64i0.10961.

Somanna, K., Somasekhara Reddy, T., & Sambasiva Rao, M. (2016). Geomorphology and evolution of the modern Mahanadi Delta using remote sensing data. *International Journal of Science and Research, 5*(2), 1329–1335.

Srivastava, S. K., Srivastava, R. C., Sethi, R. R., Kumar, A., & Nayak, A. K. (2014). Accelerating groundwater and energy use for agricultural growth in Odisha: Technological and policy issues. *Agricultural Economics Research Review, 27*(2), 259–270.

Sundaray, S. K., Panda, U. C., Nayak, B. B., & Bhatta, D. (2006). Multivariate statistical techniques for the evaluation of spatial and temporal variations in water quality of the Mahanadi river–estuarine system (India)—A case study. *Environmental Geochemistry and Health, 28*(4), 317–330. https://doi.org/10.1007/s10653-005-9001-5.

Szabo, S., Nicholls, R. J., Neumann, B., Renaud, F. G., Matthews, Z., Sebesvari, Z., et al. (2016). Making SDGs work for climate change hotspots. *Environment: Science and Policy for Sustainable Development, 58*(6), 24–33. https://doi.org/10.1080/00139157.2016.1209016.

The Economic Times. (2015, December 9). Odisha third top investment destination: Assocham. https://economictimes.indiatimes.com/news/economy/finance/odisha-third-top-investment-destination-assocham/articleshow/50108221.cms. Last accessed 7 November 2018.

Tompkins, E. L., Suckall, N., Vincent, K., Rahman, R., Mensah, A., Ghosh, A., et al. (2017). *Observed adaptation in deltas* (Deltas, Vulnerability and Climate Change: Migration and Adaptation [DECCMA] Working Paper). Southampton, UK: University of Southampton. http://www.deccma.com/deccma/Working_Papers/. Last accessed 7 November 2018.

Tumbe, C. (2011). *Remittances in India: Facts and issues* (IIM Bangalore Research Paper No. 331). Social Science Research network (SSRN). https://ssrn.com/abstract=2122689. Last accessed 10 October 2018.

Unnikrishnan, A. S., Ramesh Kumar, M. R., & Sindhu, B. (2011). Tropical cyclones in the Bay of Bengal and extreme sea-level projections along the east coast of India in a future climate scenario. *Current Science, 101*(3), 327–331.

Whitehead, P. G., Barbour, E., Futter, M. N., Sarkar, S., Rodda, H., Caesar, J., et al. (2015). Impacts of climate change and socio-economic scenarios on flow and water quality of the Ganges, Brahmaputra and Meghna (GBM) river systems: Low flow and flood statistics. *Environmental Science: Processes Impacts, 17*(6), 1057–1069. https://doi.org/10.1039/C4EM00619D.

WRIS. (2011). *River basins.* Water Resource Information System of India http://india-wris.nrsc.gov.in/wrpinfo/index.php?title=Basins. Last accessed 5 November 2018.

WRIS. (2014). *Dams in Mahanadi.* Water Resource Information System of India. http://india-wris.nrsc.gov.in/wrpinfo/index.php?title=Dams_in_Mahanadi. Last accessed 25 October 2018.

4

The Volta Delta, Ghana: Challenges in an African Setting

Samuel Nii Ardey Codjoe, Kwasi Appeaning Addo, Cynthia Addoquaye Tagoe, Benjamin Kofi Nyarko, Francisca Martey, Winfred A. Nelson, Philip-Neri Jayson-Quashigah, D. Yaw Atiglo, Prince Osei-Wusu Adjei, Kirk Anderson, Adelina Mensah, Patrick K. Ofori-Danson, Barnabas Akurigo Amisigo, Jennifer Ayamga, Emmanuel Ekow Asmah, Joseph Kwadwo Asenso, Gertrude Owusu, Ruth Maku Quaye and Mumuni Abu

4.1 The Volta Delta: Evolution and Biophysical Characteristics

Located in south-eastern Ghana between Longitudes 0° 40′ E and 1° 10′ E and Latitudes 5° 25′ N and 6° 20′ N, the Volta Delta of Ghana covers an area of about 4562 km^2 (Appeaning Addo et al. 2018)

S. N. A. Codjoe (✉) · D. Y. Atiglo · K. Anderson · G. Owusu · R. M. Quaye · M. Abu
Regional Institute for Population Studies, University of Ghana, Legon-Accra, Ghana
e-mail: scodjoe@ug.edu.gh

K. Appeaning Addo · P.-N. Jayson-Quashigah · P. K. Ofori-Danson
Department of Marine and Fisheries Sciences, Institute for Environment and Sanitation Studies, University of Ghana, Legon-Accra, Ghana

R. J. Nicholls et al. (eds.), *Deltas in the Anthropocene*,
https://doi.org/10.1007/978-3-030-23517-8_4

(Fig. 4.1). The Volta Delta lies within the Keta Basin, which is one of several fault-controlled sedimentary basins in West Africa (Appeaning Addo et al. 2018). The basin is underlain by acid and basic gneisses and schists of the Dahomeyan system with outcrop on the northern fringes of the basin. The soil underlying the Keta Basin is soft, highly compressible organic or inorganic clays overlaying fine sand to great depth (Kumapley 1989).

The delta plain is almost flat, featureless and descends gradually from inland to the Gulf of Guinea. Several thousand years ago, the river mouth of the delta was located further east but migrated westward (Nairn et al. 1999). The repositioning of the river mouth, decrease in sediment supply and reworking of the Holocene delta plain have resulted in realignment of the delta front east of the present river mouth (Anthony 2015).

The Volta Delta falls within the south-eastern coastal plains climatic zone. Climatic conditions are influenced by two air masses, the dry north-east trade winds and the moist southwest monsoon winds, which

---

K. Appeaning Addo · A. Mensah · J. Ayamga
Institute for Environment and Sanitation Studies, University of Ghana, Legon-Accra, Ghana

C. Addoquaye Tagoe
Institute of Statistical, Social and Economic Research, University of Ghana, Legon-Accra, Ghana

B. K. Nyarko
Geography and Regional Planning, University of Cape Coast, Cape Coast, Ghana

F. Martey
Research Department, Ghana Meteorological Agency, Legon-Accra, Ghana

W. A. Nelson
National Development Planning Commission, Accra, Ghana

P. O.-W. Adjei
Department of Geography and Rural Development, Kwame Nkrumah University of Science and Technology, Kumasi, Ghana

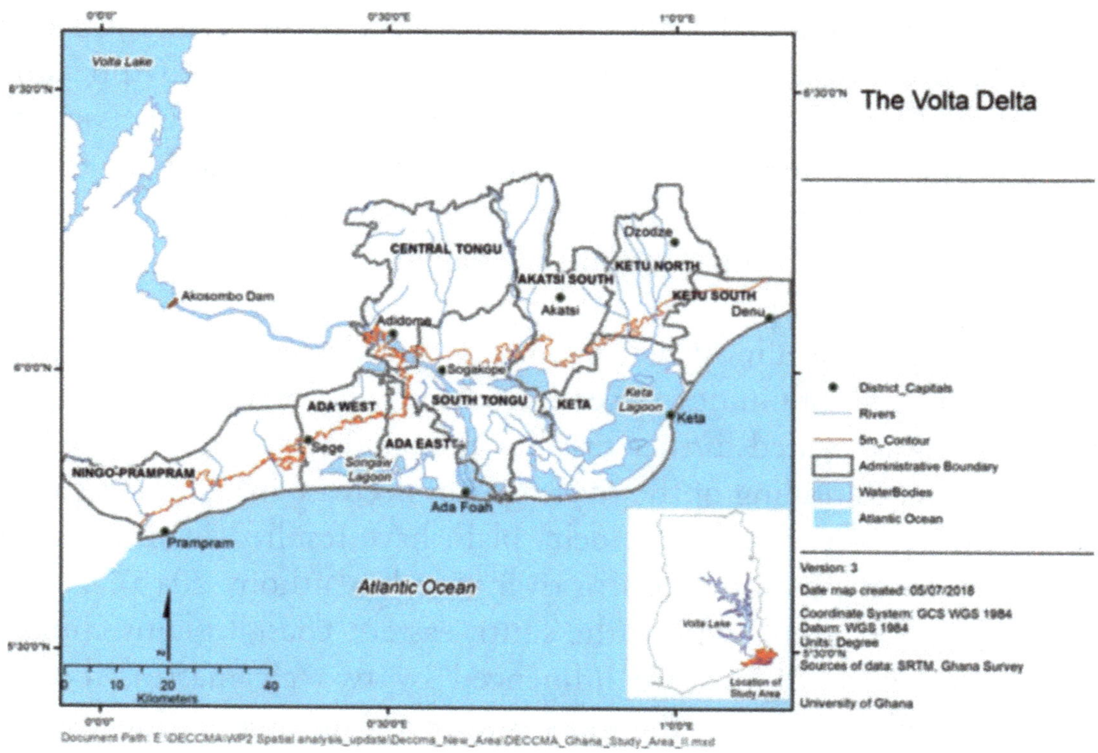

Fig. 4.1 Map of the Volta Delta area depicting the 5 m contour boundary of the delta and the nine administrative districts partly or wholly within the delta

produce a double maximum rainfall pattern (Gampson et al. 2017). The major rainy season falls between March and July, and the minor rainy season is between August and November. Mean annual rainfall varies between 146 and 750 mm between years and increases from south of the delta to its north. The dry season occurs from November to February

B. A. Amisigo
Water Research Institute, Council for Scientific and Industrial Research, Accra, Ghana

E. Ekow Asmah
Department of Economics,
University of Cape Coast, Cape Coast, Ghana

J. K. Asenso
Ministry of Finance, Government of Ghana, Accra, Ghana

and it is characterised by the north-east trade winds (harmattan). The mean temperature does not fall below 25 °C (Awadzi et al. 2008).

Ocean wave action significantly shapes the delta front. Swell waves of moderate to high energy with an average height of about 1.4 metres (m) and a long period of about 11 seconds (s) approach the shoreline unimpeded from the south to south west (Almar et al. 2015). The consistent wave action generates significant longshore currents which transport sediment eastwards, causing one of the highest rates of annual unidirectional longshore sand drift in the world (1–1.5 × 106 m^3/yr) (Nairn et al. 1999). Tides are semi-diurnal with a tidal range of about 1 m and generate weak currents which have limited effect on the shoreline morphology (Appeaning Addo et al. 2008).

The Volta River is one of the main sources of sediment supply to the Gulf of Guinea but its activities have been interfered with by anthropogenic activities. The river's discharge varied between 1000 m^3/s in the dry season and over 6000 m^3/s in the wet season before the Akosombo Dam was completed in 1965 (Anthony et al. 2015). Runoff before dam construction was 87.5 mm/yr and more varied than the post-dam period at 73.5 mm/yr (Oguntunde et al. 2006). The natural flooding patterns of the area have changed due to the controlled flow of water. In addition, annual sediment transport has drastically reduced by dam construction to only a fraction of the original transport with no peaks in flow discharge (Bollen et al. 2011) (see Chapter 5). The delta comprises extensive swamps, interspersed with short grassland mangrove areas, mainly red mangrove, and savannah woodland (Manson et al. 2013).

4.2 Demographic and Socio-Economic Characteristics

Table 4.1 presents a comparison of some key demographic characteristics of the Volta Delta and Ghana. The population of the Volta Delta as defined in Fig. 4.1 was about 850,000 in 2010 constituting about 4% of the total population of Ghana (24.6 million). The population structure has a broad base with people below 15 years constituting about

Table 4.1 Key demographic characteristics of the Volta Delta and Ghana (Data from the 2010 Ghana National Population and Housing Census [GSS 2013])

Population dynamic	Volta Delta	Ghana
Population	856,050	24,658,823
Proportion of population below 15 years	38.0%	38.3%
Proportion of population aged 65 years and above	7.1%	4.7%
Dependency ratio	81	76
Sex ratio	87.0	95.2
Proportion of female-headed households	41%	34%
Population density (persons per km^2)	151	103
Total fertility rate	3.6	4.0
Population growth rate	1.6%	2.1%
Proportion urban	33%	52%

a third of the total population of the delta (38%) similar to Ghana nationally (38.3%). Further, the population aged 65 years and above constitutes 7% of the delta population which is much higher than the national level of just under 5%. As such, the dependency ratio (84 dependent-age population per 100 population of working-age [15–64]) in the Volta Delta is higher than the national ratio (76 per 100). While the sex ratio in the delta is lower (88 males per 100 females) compared to the national average (95.2 males per 100 females), the proportion of female-headed households is higher in the Volta Delta (45%) compared to the national average (34%) (Appeaning Addo et al. 2018). A lower sex ratio coupled with a higher proportion of female-headed households in the Volta Delta can be attributed to high out-migration of males (Atiglo and Codjoe 2015).

The population density of the Volta Delta is 151 persons per km^2 compared to 103 persons per km^2 in Ghana. Both the Total Fertility Rate (3.6) and annual population growth rate (1.6%) are lower in the Volta Delta compared with the national averages of 4.0 and 2.1% respectively. About two-thirds of the delta population is rural, however, it is projected that more than half of the population of the Volta Delta will be living in urban areas by 2035 through migration. Increasing population growth and urbanisation have altered the land cover, topography and land use in the delta region, and vegetated land has been converted to agricultural use and settlements (Appeaning Addo 2015).

Regarding socio-economic characteristics, the dominant types of dwelling are the compound house and separate structures, and slightly over 60% of dwelling units are owned by households (GSS 2013). Outer wall materials of buildings are mainly concrete, and cement block. Biomass (wood, charcoal, sawdust, etc.) constitutes the main source of fuel for cooking and kerosene the main source for lighting. However, considering the rapid rural electrification that has occurred in the country over the past twenty years, it is expected that more households will be connected to the national electricity grid (Kumi 2017). This will increase the number of households that use electricity for cooking and lighting. While the main source of water for drinking and domestic use is pipe-borne, unimproved sources, including groundwater from wells and open water sources (30%) are quite common.

The two main ethnic groups in the Volta Delta are Ewe and Ga-Dangme and both are patrilineal. The dominant religion in the Volta Delta is Christianity (72%), followed by Traditional African Religion (22%) and Islam (3%). Illiteracy rate for the population aged 15 years and above is about 30%, with higher illiteracy rates for females compared to males (GSS 2013).

The Volta Delta has a diverse economic system with different but integrated sectors, i.e., agriculture including livestock rearing and fisheries, salt and sand mining, construction, trade, transport and tourism (Codjoe et al. 2017). Agriculture-related activities are very common and important for livelihoods in the delta. The crop production sector employs a higher proportion of females than males (Barry et al. 2005). Females mainly engage in processing and selling of fish and farm produce, while males mostly engage in fishing and crop production (Ayivor and Kufogbe 2001). It is, however, largely small-scale and characterised by unsophisticated technologies or irrigation systems. Coconut, for instance, was a major cash crop in the delta in the nineteenth century. However, in the 1930s, production was affected by the Cape St Paul Wilt disease leading to the collapse of the industry (Eziashi and Omamor 2010). In the last two decades, the services sector, particularly trade, transport and small and medium scale manufacturing, has seen a major boost. In recent times, aquaculture farms have been established in the River Volta and the Keta Lagoon and this industry has huge potential to supplement declining marine fish stocks in the future (Amponsah et al. 2015; Lauria et al. 2018).

Other emerging economic activities include charcoal burning which involves cutting of trees (Akrasi 2005) and salt production (Barry et al. 2005). Furthermore, coastal sand mining which mainly feeds the booming construction industry in the nearby cities of Accra and Tema, despite being illegal, is a common activity (Mensah 2002; Anim et al. 2013; Wiafe et al. 2013; Jonah et al. 2015). Indeed, this practice negatively influences the sediment budget and has contributed significantly to increased erosion along the coast (Appeaning Addo 2015).

Finally, the Volta Delta has very important tourist attractions such as marine turtle breeding sites located in the estuary at Totope, Lolonya, Akplabanya and Kewuse, bird watching on the Songhor and Keta Lagoons (designated wetlands and Ramsar sites which provide sanctuary for about 80% of migratory birds that transit in Ghana), fetish shrines, sacred groves and traditional festivals. Resorts located on the Volta River provide water recreational activities and attract migrants to work in the tourism industry (Codjoe et al. 2017).

Short to medium term economic projections of the delta show that the manufacturing sector could outpace the agriculture and services sectors, partly due to impacts of environmental changes on agricultural production as well as prospects for future oil production (Adjei et al. 2016). Currently, the delta remains a net importer of goods and services as economic activities are inadequate to satisfy domestic and external demands.

4.3 Biophysical and Socio-Economic Drivers of Change

The Volta Delta is a dynamic and rich environment that is constantly changing in time and space (Dada et al. 2016). The delta, formed some thousands of years ago from sediments deposited along the mouth of the River Volta, and has constantly undergone changes since the early Holocene (Nairn et al. 1999). Natural and anthropogenic factors combine in a complex system to drive such change. In most instances, the anthropogenic factors tend to exacerbate the natural effects. It is to be noted, however, that anthropogenic interventions have not always been detrimental to the environment as laws and policies have been enacted overtime to protect the natural resources in the delta area.

Climate change-related events such as rainfall variability, marine and riverine flooding, drought, sea-level rise, storm surges and increased temperature are some natural drivers of change in the biophysical conditions of the delta system (Appeaning Addo et al. 2018). Human-induced changes occur mainly from interruptions in the hydrology, land use and the landscape. The onset of the rainy season has changed due to climate variability. However, increased rainfall intensity in the rainy season causes frequent flooding in the Volta Delta. The floods degrade the environment, threaten lives, destroy properties, and result in displacement of households. Energetic swell waves, increasing sea-level rise and storm surge aided by relatively low topography also facilitate coastal flooding. The delta coast is dynamic with high crests underlain by soft rocks which naturally expose the coast from Prampram to Aflao to flooding, erosion and shoreline recession (Ly 1980). Erosion in the Volta Delta was first reported in 1929 but posited to have existed since the 1860s, particularly in Keta (Nairn et al. 1999).

A critical characteristic of the Anthropocene delta was the construction of the Akosombo and Kpong Dams in 1964 and 1982, respectively, which drastically reduced sediment supply to the coast. This exposed it to severe wave action and erosion resulting in accelerated shoreline retreat, reduction in fresh water and fish supply down-stream (Ly 1980; Nairn et al. 1999; Tsikata 2006) and also the introduction of water hyacinths that affect aquatic life in the Volta River (Gyau-Boakye 2001). Furthermore, it has been documented that schistosomiasis infections became a public health issue in Ghana after the construction of the Akosombo Dam in the 1960s, which created extensive areas suitable for the breeding of the schistosoma host snails (Paperna 1970). The prevalence of urinary schistosomiasis was below 10% among communities living along the river before damming in 1964 and reached as high as 80–90% after damming by 1971 (Lavoipierre 1973; Barry et al. 2005).

Additionally, the construction of the Tema Harbour in 1955 caused sea wave diffraction on the land along the east coast of Ghana, causing massive erosion (Ly 1980; Tsidzi and Kumapley 2001). Erosion rates of about 4 m per year before the construction of the dams increased to about 8 m per year post-dam construction (Ly 1980). To address some impacts of erosion, the Keta and Ada Sea Defence Projects were undertaken in 2001 and 2013, respectively, by the central government. These were preceded by attempts to protect the shoreline by communities in Keta and its environs, led by their traditional leaders and the government, since 1923. These earlier defence structures from colonial times were constructed with weak local materials which could not withstand the strong sea waves (Akyeampong 2002; van der Linden et al. 2013). After Ghana's independence in 1957, the government undertook a coastal protection project in 1960 using steel sheets to protect about 1,600 m of the Keta Township but these also corroded rapidly (van der Linden et al. 2013).

By 1996, the rate of erosion and flooding had increased, with more than half of Keta and its surrounding towns under water (Ile et al. 2014) displacing more than 10,000 people in communities within Keta and leading to losses in the millions of dollars (Oteng-Ababio et al. 2011; Danquah et al. 2014). The central government therefore undertook a sea defence project to construct a causeway across the Keta lagoon for the coastal highway, reclaim lands lost to the sea and construct houses for resettlement of displaced people (Danquah et al. 2014). Between 2001 and 2004, six groynes were constructed within the Keta area to prevent erosion and to control flooding of buildings between the Keta Lagoon and the sea (Boateng 2009). The groynes were about 190 m in length and 750 m apart (Nairn and Dibajnia 2004). In 2011, the Government of Ghana began the construction of a 30 km sea defence wall at Ada to protect communities against wave action (Anim et al. 2013). The defence structures have had two effects; accretion of the up-drift side and increased erosion of the down-drift side of the shoreline (Wiafe et al. 2013; Appeaning Addo 2015).

In recent times, groundwater extraction for irrigated farming practices which has the potential for enhanced subsidence of the delta at a rate of about 1 mm per annum as pertains in other deltas globally

(Kortatsi et al. 2005; Appeaning Addo et al. 2018), mangrove harvesting which may cause erosion and flooding (Anim et al. 2013), prospecting for oil and gas which is expected to increase the rate of subsidence (Setordzi and Nyavor 2015) and coastal sand mining with the potential of reducing sediment budget and thus increase erosion (Appeaning Addo 2015; Appeaning Addo et al. 2018) are being widely practiced.

4.4 Adaptation to Climate and Non-climate Change

Households and communities in the Volta Delta employ both autonomous and planned adaptation strategies to respond to the multiple threats of climatic and non-climatic stressors. Drought and flooding (both riverine and coastal) are the most common pressures on traditional livelihoods, in addition to land degradation, land-use changes, and increasing population demands on natural resources.

In the delta, as in most parts of the country, the prioritised sectors for adaptation are agriculture, water resources and disaster risk reduction. In the post-dam Anthropocene era, the government and non-governmental organisations have implemented a number of projects and programmes to support communities to deal with the inundation of farmlands and settlements, loss of property, salinisation of groundwater, low agricultural productivity, coastal and riverine erosion, water shortage and increased incidence of water and sanitation-related diseases. Planned adaptation activities have mainly focused on implementing change with improved technologies, building capacity (e.g., climate change awareness and adaptation governance training programmes), providing alternative livelihoods for improving food security, minimising the impacts of flooding on communities through disaster risk reduction and rural-urban development actions, improving ecological functions and services for providing sustainable access to natural resources (such as mangrove regeneration and tree planting), and improving access to water.

Evidence from a household survey carried out in 2016 (see DECCMA 2018) shows that, at the community level, about three-quarters of households have undertaken one or multiple forms of autonomous adaptation strategies. The main adaptation strategies undertaken by households include modification of labour utilisation which involves the use of hired labour and/or women taking up work outside the home. Also, modifications to housing structures or moving houses entirely are common. As noted earlier, houses are usually made of weak materials and may require regular modification or complete abandonment. Financial capital investments in the form of loans, insurance and joining cooperatives are also common. Among these, insurance is the least employed strategy (not shown). Migration does not feature as a major adaptation strategy among households.

Farming and fishing related adaptation strategies constitute the least used strategies in the delta as a whole. Receiving training in new fishing or farming skills is the least common adaptation strategy in the delta. However, among farming households only, purchasing farm equipment is most common (Fig. 4.2).

The most successful adaptation options for most households are strategies that increase their resilience, reduce disaster impacts and improve

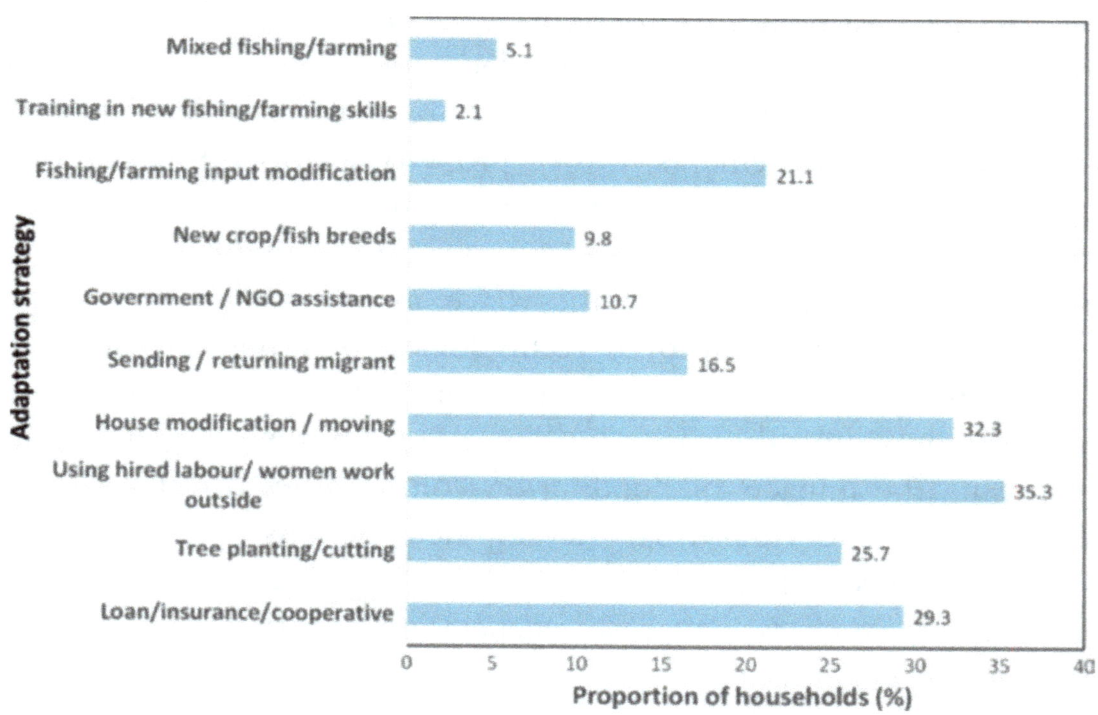

Fig. 4.2 Adaptation strategies by household adaptation strategy in the Volta Delta

living standards by improving their overall skills and capacity for improved livelihood options. Some indications of barriers to successful adaption relate to project cost implications, the ability of targeted communities/individuals to participate or sustain participation, conflicts between original landowners and related communities over land ownership, and concerns about future loss of biodiversity and downstream pollution from the establishment of commercial agricultural farms. Other barriers include concerns over impacts of constructed sea defence structures on the movement and nesting patterns of sea turtles, the threat of non-indigenous tree species used as buffers along the banks of the Volta River becoming invasive species, and possible interruptions to water supply due to more communities getting connected to the same source.

There are positive influences on the sustainability of adaptation actions due to shared ownership and commitment by the local community through active participation in planning, design and implementation; commitment and financial support (e.g., subsidies) demonstrated by the government; self-generated funds to support action; and the use of cost effective and efficient technologies. Adaptation options that have built in sustainability and ecologically sensitive components are also more likely to be effective in the long-term for reducing disaster risk and vulnerability of the communities, and increasing resilience of the system.

Finally, gendered issues in adaptation were not prominent since both males and females generally have similar perspectives on how effective or successful an adaptation strategy is. However, introducing new breeds in fishing ponds, converting mono-cropped land to mixed farming, employing irrigation farming techniques and relocating to a new house are ranked higher by females, and males rank receiving governmental and NGO intervention and being trained with new skills higher compared to females.

4.5 Migration and Resettlement

The first recorded mass migration into the Volta Delta was in the sixteenth century by the Ga-Dangme and Ewe ethnic groups. Ever since this period, there has been migration to and from the delta although out-migration has been predominant. In the Anthropocene, the Volta Delta is a net migrant sending area, although some districts (South Tongu, Ketu South, Akatsi South, Ada East, Ada West and Prampram) are net-migrant receiving areas. Based on the results of the survey carried out for this research, the main drivers of migration from the delta are economic (employment opportunities), education and family reunion; very few individuals cite direct environmental factors as the main reason for migrating. Furthermore, higher proportions of males, mainly youth, migrate due to economic reasons, whilst more females migrate to reunite with a family or for education. It should be noted, however, that some of the economic or livelihood drivers of migration may be directly linked to climate and/or environmental change.

People who migrate for economic reasons are usually never married, cohabiting or separated. Movements from the delta area are usually permanent and there is continuous mobility from the area. Respondents in male-headed households have a higher proportion of members who intend to migrate compared to female-headed households. Although more than a third of the respondents (42%) in the Volta Delta are of the view that migration is disruptive to individuals and the household, there is a general perception that migration is helpful. The common perception is that migration improves the socio-economic conditions of individuals and households and this feedback may continue to increase future migration from the delta, if economic opportunities are not enhanced in situ.

Some biophysical factors are associated with out-migration from the region. For example, there is already high impact of climate change on farming and fishing activities in the area. Land used for crop and vegetable cultivation and landing sites for fishers have been lost to flooding and erosion, and fresh water fishing and farming in or near the lagoon have also been affected by salinisation (Codjoe et al. 2017). In addition, the construction of the sea defence walls has not attracted people back into areas destroyed by coastal erosion or inundation due to lost livelihoods and rare economic opportunities. It is estimated that future accelerated environmental change due to the impacts of coastal erosion and flooding will lead to mass displacements and high out-migration from affected areas.

Successive governments have considered resettling highly vulnerable communities in the delta. Prior to independence, earlier attempts to resettle communities from the Keta area failed because of the communities' attachment to their place and main livelihood activity, i.e., fishing (Akyeampong 2002). In 1996, the government undertook a comprehensive resettlement scheme for three communities in Keta—Adzido, Vodza and Kedzi (Akyeampong 2002; Afram et al. 2015). According to Afram et al. (2015), resettled communities are generally happy with the housing facilities provided but are dissatisfied with the living arrangements. In particular, people prefer to live close to their extended families and to have spaces around their homes for gatherings such as funerals and marriage ceremonies. Not incorporating these cultural and family contexts result in loss of family and cultural ties, violation of people's fundamental human rights as well as apathy for community cohesion (Danquah et al. 2014).

4.6 Delta Management and Policy

As shown in Table 4.2, pre-colonial management of the Volta Delta environment was primarily based on traditional/customary laws which included sanctions, taboos, cultural norms, beliefs, values, codes of conduct, etc. that were handed down over generations by oral tradition, teaching or imitation (Kuupole and Botchway 2010). Management of the delta was vested in chiefs and traditional leaders who ensured that all natural resources inherited from their predecessors were sustained for the next generation (Opoku-Agyemang 2001a). For example, there were customary rules and beliefs which prohibited fishing activities on some days and farming close to the water body or harvesting mangroves, and there were demarcated sacred groves in order to protect the serene and fragile environment (Sarpong 2004; Atampugre et al. 2016).

There is no explicit national policy devoted entirely to the Volta Delta. However, various policies contain focus areas that apply to protecting and conserving some delta features. During colonial and post-colonial times, formal rules in the form of laws, policies, protocols

Table 4.2 Laws, policies, conventions and ordinances and their impacts on the Volta Delta

Year/period	Policy type	Level	Impact	Source
Pre-colonial era	Traditional/customary laws i.e., taboos, restrictions etc.	Local community	Mangrove protection Fish conservation	Kuupole and Botchway (2010) Opoku-Agyemang (2001a)
Colonial era	River Ordinance, Forests Ordinance	National	Protection of rivers, forests and mangroves	Opoku-Agyemang (2001b)
1971	Convention on Wetlands of International Importance Especially as Water Fowl Habitats: Ramsar	International	Wetland protection	Opoku-Agyemang (2001b)
1981	Convention for the Cooperation in the Protection and Development of the Marine and Coastal Environment of the West and Central African Region (Abidjan Convention)	International (regional)		Opoku-Agyemang (2001b)
1985	United Nations Convention on the Law of the Sea	International		Opoku-Agyemang (2001b)
1996	Integrated Water Resource Management Policy	National	Groundwater management	
1999	National Wetland Strategy	National	Wetland and wild-life protection	
2001	Water Use Regulation Policy LI 1962	National	Water management	

(continued)

Table 4.2 (continued)

Year/period	Policy type	Level	Impact	Source
2002	Declaration on improved management of natural resources of the Volta Basin	International (regional)		
2006	Drilling for Water and Groundwater Development Regulation Policy LI 1827	National	Groundwater management	
2007	National Water Policy	National	Water management	MWWH (2007)
2011	National Irrigation Policy, Strategies and Measures Document	National	Irrigation management	MFA (2011)
2011	National Buffer Zone Policy	National	Protection of buffer zones	MWWH (2011)
Ongoing	National Coastal Policy	National		

and agreements were established to ensure delta management. The earliest documented law on the Volta River Basin in Ghana was the River Ordinance of 1903 (Opoku-Agyemang 2001b) to manage the use of some rivers in the country, including the Volta, and vested powers of control over the river in the colonial governments.

National policies with focus areas that may help protect the delta include the National Water Policy; the National Wetlands Policy; the Tourism Development Policy, Land Management Policy; National Environmental Policy; Energy Policy; Minerals Policy and the Wildlife Conservation Policy. A national coastal policy is currently under development. International protocols that apply to the delta include the Convention on Wetlands of International Importance Especially as Water Fowl Habitats: Ramsar Convention (1971); International Covenant on Economic, Social and Cultural Rights (2000); United Nations Convention on the Law of the Sea (1985); Volta Basin Declaration on improved management of the natural resources of the Volta Basin (2002); Convention for the Cooperation in the Protection and Development of the Marine and Coastal Environment of the West and Central African Region (Abidjan Convention) (1981).

4.7 Conclusion

There has been a transitioning in the Volta Delta from the Holocene to the Anthropocene. It is evident that human activities have greatly altered the physical characteristics in the delta area and the built environment has further reshaped the nature of human settlement, livelihoods and movement. The construction of the Akosombo and Kpong Dams has had tremendous impact, not only on the geophysical characteristics of the delta, but also on the socio-economic adjustments to population livelihoods (see also Chapter 8). The dams have interfered with sedimentation process and repositioning of the river mouth along the Gulf of Guinea and also reduced the amount of freshwater that flows into the sea. Further, the fish population that flowed from upstream has significantly dropped. Thus, there has been outmigration from the delta area to upstream areas for freshwater fishing.

Similarly, the construction of a fishing harbour in Tema, west of the delta has impacted the rates of erosion as well as livelihoods in the delta region. The harbour and an associated industrial city serve as an attractive hub for migrants from the delta whose main fishing-based livelihoods have been adversely impacted by the construction of the dams. Migration is dominated by males leading to a higher proportion of females in the delta.

Currently, the construction of groynes along the coasts of Keta and Ada have resulted in significant rates of accretion and reshaping of the morphology along the shoreline. The Keta sea defence project and accompanying resettlement scheme continue to have some impact on the physical nature of the delta as well as human mobility. The construction of the revetment in Atokor in the Keta Municipality has further enabled the reconstruction of the previously destroyed road and opened up the area for commercial activities. Finally, human settlements, land use and economic activities continue to reshape the land cover and biosphere of the delta area.

References

Adjei, P. O.-W., Ofori-Danson, P. K., Asenso, J. K., & Amponsah, S. K. (2016). *Biophysical and socioeconomic state of the Volta Delta region of Ghana from the perspectives of gender and spatial relations* (Deltas, Vulnerability and Climate Change: Migration and Adaptation [DECCMA] Working Paper). Southampton, UK: DECCMA Consortium. https://generic.wordpress.soton.ac.uk/deccma/resources/working-papers/. Last accessed 5 December 2016.

Afram, S. O., Kwofie, T. E., & Attipoe, J. (2015, March 25–26). *The influence of beneficiary participation in resettlement schemes in Ghana. A case study of the Keta Basin Sea Defence Resettlement Project.* Proceedings of the 4th International Conference on infrastructure Development in Africa, Kumasi, Ghana.

Akrasi, S. A. (2005). The assessment of suspended sediment inputs to Volta Lake. *Lakes & Reservoirs: Science, Policy and Management for Sustainable Use, 10*(3), 179–186. https://doi.org/10.1111/j.1440-1770.2005.00272.x.

Akyeampong, E. (2002). *Between the sea and the lagoon: An eco-social history of the Anlo of southeastern Ghana, c. 1850 to recent times.* Martlesham, UK: James Currey.

Almar, R., Kestenare, E., Reyns, J., Jouanno, J., Anthony, E. J., Laibi, R., et al. (2015). Response of the Bight of Benin (Gulf of Guinea, West Africa) coastline to anthropogenic and natural forcing, Part1: Wave climate variability and impacts on the longshore sediment transport. *Continental Shelf Research, 110,* 48–59. https://doi.org/10.1016/j.csr.2015.09.020.

Amponsah, S. K., Danson, P. O., Nunoo, F. K. E., & Lamptey, A. M. (2015). *Assessment of security of coastal fishing in Ghana from the perspectives of safety, poverty and catches* (Master's thesis). University of Ghana, Accra, Ghana.

Anim, D. O., Nkrumah, P. N., & David, N. M. (2013). A rapid overview of coastal erosion in Ghana. *International Journal of Scientific & Engineering Research, 4*(2), 1–7.

Anthony, E. J. (2015). Patterns of sand spit development and their management implications on deltaic, drift-aligned coasts: The cases of the Senegal and Volta River delta spits, West Africa. In G. Randazzo, D. Jackson, & A. Cooper (Eds.), *Sand and gravel spits* (pp. 21–36). Heidelberg, Germany: Springer.

Anthony, E. J., Brunier, G., Besset, M., Goichot, M., Dussouillez, P., & Nguyen, V. L. (2015). Linking rapid erosion of the Mekong River delta to human activities. *Scientific Reports, 5.* https://doi.org/10.1038/srep14745.

Appeaning Addo, K. (2015). Monitoring sea level rise-induced hazards along the coast of Accra in Ghana. *Natural Hazards, 78*(2), 1293–1307. https://doi.org/10.1007/s11069-015-1771-1.

Appeaning Addo, K., Nicholls, R. J., Codjoe, S. N. A., & Abu, M. (2018). A biophysical and socioeconomic review of the Volta Delta, Ghana. *Journal of Coastal Research.* https://doi.org/10.2112/JCOASTRES-D-17-00129.1.

Appeaning Addo, K., Walkden, M., & Mills, J. P. (2008). Detection, measurement and prediction of shoreline recession in Accra, Ghana. *ISPRS Journal of Photogrammetry and Remote Sensing, 63*(5), 543–558. https://doi.org/10.1016/j.isprsjprs.2008.04.001.

Atampugre, G., Botchway, D.-V. N. Y. M., Esia-Donkoh, K., & Kendie, S. (2016). Ecological modernization and water resource management: A critique of institutional transitions in Ghana. *GeoJournal, 81*(3), 367–378. https://doi.org/10.1007/s10708-015-9623-9.

Atiglo, D. Y., & Codjoe, S. N. A. (2015). Migration in the volta delta: A review of relevant literature (Deltas, Vulnerability and Climate Change: Migration and Adapation [DECCMA] Working Paper). IDRC Project Number 107642. www.deccma.com.

Awadzi, T. W., Ahiabor, E., & Breuning-Madsen, H. (2008). The soil-land use system in a sand spit area in the semi-arid coastal savanna region of Ghana:

Development, sustainability and threats. *West African Journal of Ecology, 13,* 132–143. https://doi.org/10.4314/wajae.v13i1.40573.

Ayivor, J. S., & Kufogbe, S. K. (2001). Post-dam agro-ecological challanges of the lower volta basin in Ghana. *Bulletin of the Ghana Geographical Association, 23*(1), 88–102.

Barry, B., Obuobie, E., Andreini, M., Andah, W., & Pluquet, M. (2005). *The Volta river basin. Comparative study of river basin development and management* (Report), p. 51. Colombo, Sri Lanka: International Water Management Institute (IWMI) and Comprehensive Assessment of Water Management in Agriculture (CAWMA). http://www.iwmi.cgiar.org/assessment/files_new/research_projects/river_basin_development_and_management/VoltaRiverBasin_Boubacar.pdf. Last accessed 14 August 2018.

Boateng, I. (2009, May 3–8). *Development of integrated shoreline management planning: A case study of Keta.* Federation of International Surveyors Working Week. Eilat, Israel. https://www.fig.net/resources/proceedings/fig_proceedings/fig2009/papers/ts04e/ts04e_boateng_3463.pdf. Last accessed 14 August 2018.

Bollen, M., Trouw, K., Lerouge, F., Gruwez, V., Bolle, A., Hoffman, B., et al. (2011). *Design of a coastal protection scheme for Ada at the Volta River mouth (Ghana).* Proceedings of 32nd Conference on Coastal Engineering, 2010. Shanghai, China. International Conference on Coastal Engineering (ICCE). https://doi.org/10.9753/icce.v32.management.36.

Codjoe, S. N. A., Nyamedor, F. H., Sward, J., & Dovie, D. B. (2017). Environmental hazard and migration intentions in a coastal area in Ghana: A case of sea flooding. *Population and Environment, 39*(2), 128–146. https://doi.org/10.1007/s11111-017-0284-0.

Dada, O. A., Li, G., Qiao, L., Ma, Y., Ding, D., Xu, J., et al. (2016). Response of waves and coastline evolution to climate variability off the Niger Delta coast during the past 110 years. *Journal of Marine Systems, 160,* 64–80. https://doi.org/10.1016/j.jmarsys.2016.04.005.

Danquah, J. A., Attippoe, J. A., & Ankrah, J. S. (2014). Assessment of residential satisfaction in the resettlement towns of the Keta Basin in Ghana. *International Journal of Civil Engineering, Construction and Estate Management, 2*(3), 26–45.

DECCMA. (2018). *Climate change, migration and adaptation in deltas: Key findings from the DECCMA project* (Deltas, Vulnerability and Climate Change: Migration and Adaptation [DECCMA] Report). Southampton, UK: DECCMA Consortium. https://www.preventionweb.net/publications/view/61576. Last accessed 27 November 2018.

Eziashi, E., & Omamor, I. (2010). Lethal yellowing disease of the coconut palms (cocos nucifera l.): An overview of the crises. *African Journal of Biotechnology, 9*(54), 9122–9127.

Gampson, E. K., Nartey, V. K., Golow, A. A., Akiti, T. T., Sarfo, M. A., Salifu, M., et al. (2017). Physical and isotopic characteristics in peri-urban landscapes: A case study at the lower Volta River Basin. *Ghana. Applied Water Science, 7*(2), 729–744. https://doi.org/10.1007/s13201-015-0286-y.

GSS. (2013). *2010 population and housing census* (National Analytical Report). Accra, Ghana: Ghana Statistical Service. http://www.statsghana.gov.gh/docfiles/publications/2010_PHC_National_Analytical_Report.pdf. Last accessed 28 August 2018.

Gyau-Boakye, P. (2001). Environmental impacts of the Akosombo Dam and effects of climate change on the lake levels. *Environment, Development and Sustainability, 3*(1), 17–29. https://doi.org/10.1023/A:1011402116047.

Ile, I. U., Garr, E. Q., & Ukpere, W. I. (2014). Monitoring infrastructure policy reforms and rural poverty reduction in Ghana: The case of Keta Sea Defence Project. *Mediterranean Journal of Social Sciences, 5*(3), 633. https://doi.org/10.5901/mjss.2014.v5n3p633.

Jonah, F. E., Agbo, N. W., Agbeti, W., Adjei-Boateng, D., & Shimba, M. J. (2015). The ecological effects of beach sand mining in Ghana using ghost crabs (Ocypode species) as biological indicators. *Ocean and Coastal Management, 112,* 18–24. https://doi.org/10.1016/j.ocecoaman.2015.05.001.

Kortatsi, B. K., Young, E., & Mensah-Bonsu, A. (2005). Potential impact of large scale abstraction on the quality of shallow groundwater for irrigation in the Keta Strip, Ghana. *West African Journal of Applied Ecology, 8,* 1. https://doi.org/10.4314/wajae.v8i1.45780.

Kumapley, N. K. (1989). The geology and geotechnology of the Keta basin with particular reference to coastal protection. In W. J. M. van der Linden, S. A. P. L. Cloetingh, J. P. K. Kaasschieter, W. J. E. van de Graaff, J. Vandenberghe, & J. A. M. van der Gun (Eds.), *Coastal lowlands: Geology and geotechnology* (pp. 311–320). Dordrecht, The Netherlands: Springer.

Kumi, E. N. (2017). *The electricity situation in Ghana: Challenges and opportunities* (CGD Policy Paper). Washington, DC: Center for Global Development. https://www.cgdev.org/sites/default/files/electricity-situation-ghana-challenges-and-opportunities.pdf. Last accessed 13 November 2018.

Kuupole, D. D., & Botchway, D.-V. N. Y. M. (2010). *Polishing the pearls of ancient wisdom: Exploring the relevance of endogenous African knowledge systems for sustainable development in postcolonial Africa.* Faculty of Arts,

University of Cape Coast and Centre for Indigenous Knowledge and Organizational Development, Cape Coast, Ghana.

Lauria, V., Das, I., Hazra, S., Cazcarro, I., Arto, I., Kay, S., et al. (2018). Importance of fisheries for food security across three climate change vulnerable deltas. *Science of the Total Environment, 640–641,* 1566–1577. https://doi.org/10.1016/j.scitotenv.2018.06.011.

Lavoipierre, G. J. (1973). Development in economics and health. Schistosomiasis epidemic of Lake Volta. *Revue de l'infirmière, 23*(1), 51–53.

Ly, C. K. (1980). The role of the Akosombo Dam on the Volta river in causing coastal erosion in central and eastern Ghana (West Africa). *Marine Geology, 37*(3), 323–332. https://doi.org/10.1016/0025-3227(80)90108-5.

Manson, A. A. B., Appeaning Addo, K., & Mensah, A. (2013). Impacts of shoreline morphological change and sea level rise on mangroves: The case of the Keta coastal zone. *E3 Journal of Environmental Research and Management, 40*(10), 0334–0343.

Mensah, J. V. (2002). Causes and effects of coastal sand mining in Ghana. *Singapore Journal of Tropical Geography, 18*(1), 69–88. https://doi.org/10.1111/1467-9493.00005.

MFA. (2011). *National irrigation policy, strategies and measures.* Accra, Ghana: Ministry of Food and Agriculture (MFA), Government of Ghana. http://extwprlegs1.fao.org/docs/pdf/gha149500.pdf. Last accessed 13 November 2018.

MWWH. (2007). *National water policy.* Accra, Ghana: Ministry of Water, Works & Housing (MWWH), Government of Ghana. http://www.gwcl.com.gh/national_water_policy.pdf. Last accessed 13 November 2018.

MWWH. (2011). *Riparian buffer zone policy for managing freshwater resources in Ghana.* Accra, Ghana: Ministry of Water, Works & Housing (MWWH), Government of Ghana. http://extwprlegs1.fao.org/docs/pdf/gha149365.pdf. Last accessed 13 November 2018.

Nairn, R. B., & Dibajnia, M. (2004). Design and construction of a large headland system, Keta Sea Defence Project, West Africa. *Journal of Coastal Research, 33,* 294–314. Special Issue.

Nairn, R. B., MacIntosh, K. J., Hayes, M. O., Nai, G., Anthonio, S. L., & Valley, W. S. (1999). *Coastal erosion at Keta Lagoon, Ghana: Large scale solution to a large scale problem.* Proceedings of the 26th Conference on Coastal Engineering 1998, June 22–26, Copenhagen, Denmark. American society of Civil Engineers. https://doi.org/10.1061/9780784404119.242.

Oguntunde, P. G., Friesen, J., van de Giesen, N., & Savenije, H. H. G. (2006). Hydroclimatology of the Volta River Basin in West Africa: Trends and variability from 1901 to 2002. *Physics and Chemistry of the Earth, Parts A/B/C, 31*(18), 1180–1188. https://doi.org/10.1016/j.pce.2006.02.062.

Opoku-Agyemang, M. (2001a, June 27–July 1). *Shifting paradigms: Towards the integration of customary practices into the environmental law and policy in Ghana*. Proceedings of 'Securing the future: International Conference on Mining and the Environment', Skellefteå, Sweden. The Swedish Mining Association.

Opoku-Agyemang, M. (2001b, June 27–July 1). *Water Resources Commission Act and the nationalisation of water rights in Ghana*. Proceedings of 'Securing the future: International Conference on Mining and the Environment', Skellefteå, Sweden. The Swedish Mining Association.

Oteng-Ababio, M., Owusu, K., & Appeaning Addo, K. (2011). The vulnerabilities of the Ghana coast: The case of Faana-Bortianor. *JAMBA: Journal of Disaster Risk Studies, 3*(2), 429–442. https://doi.org/10.4102/jamba.v3i2.40.

Paperna, I. (1970). Study of an outbreak of schistosomiasis in the newly formed Volta lake in Ghana. *Zeitschrift fur Tropenmedizin und Parasitologie, 21*(4), 411–425.

Sarpong, G. A. (2004). *Going down the drain? Customary water law and legislative onslaught in Ghana*. Paper commissioned by Food and Agriculture Organisation as part of its studies on effect of legislation on Customary Water Rights. Rome, Italy: Food and Agriculture Organisation of the United Nations.

Setordzi, I., & Nyavor, G. (2015). Oil exploration to start soon in Keta in spite of challenges. *Joyonline news*. http://www.myjoyonline.com/news/2015/february-5th/oil-exploration-to-startsoon-in-keta-despite-challenges.php. Last accessed 14 November 2016.

Tsidzi, K. E. N., & Kumapley, N. K. (2001). Coastal erosion in Ghana: Causes and mitigation strategies. In P. G. Marinos, G. C. Koukis, G. C. Tsiambaos, & G. C. Stournaras (Eds.), *Engineering geology and the environment, volume 5* (pp. 3941–3946). Lisse, The Netherlands: A A Balkema.

Tsikata, D. (2006). *Living in the shadow of the large dams: Long term responses of downstream and lakeside communities of Ghana's Volta river Project*. African Social Studies Series. Leiden, The Netherlands: Brill.

van der Linden, W. J. M., Cloetingh, S. A. P. L., Kaasschieter, J. P. K., Vandenberghe, J., van de Graaff, W. J. E., & van der Gun, J. A. M. (Eds.). (2013). *Coastal lowlands: Geology and geotechnology*. Dordrecht, The Netherlands: Springer.

Wiafe, G., Boateng, I., Appeaning Addo, K., Quashigah, P. N. J., Ababio, S. D., & Laryea, S. (2013). *Handbook of coastal processes and management in Ghana*. Gloucester, UK: The Choir Press.

5

Fluvial Sediment Supply and Relative Sea-Level Rise

Stephen E. Darby, Kwasi Appeaning Addo, Sugata Hazra, Md. Munsur Rahman and Robert J. Nicholls

5.1 Introduction

The world's major deltas are facing a sustainability crisis (Giosan et al. 2014; Anthony et al. 2015). Of the world's 33 largest deltas, 28 are at high risk of being 'drowned' as a result of relative sea-level rise (Syvitski et al. 2009). These changes have profound implications for human habitation of these systems and suggest wide-scale abandonment, or a growing dependency on dykes, artificial drainage and polders. Given deltas' role in food production, the potential consequences for regional and global food security are readily

S. E. Darby (✉)
Geography and Environmental Science,
University of Southampton, Southampton, UK
e-mail: S.E.Darby@soton.ac.uk

K. Appeaning Addo
Department of Marine and Fisheries Sciences, Institute
for Environment and Sanitation Studies, University of Ghana,
Legon-Accra, Ghana

R. J. Nicholls et al. (eds.), *Deltas in the Anthropocene*,
https://doi.org/10.1007/978-3-030-23517-8_5

apparent: around 20% of the land within the large deltas of South and Southeast Asia will likely be lost by 2100 unless protected by dykes (Giosan et al. 2014; Smajgl et al. 2015). The challenge of securing, let alone increasing, food production is further compounded by the enhanced flood risk and waterlogging as well as the problems of saline intrusion that are also associated with sea-level rise. For these reasons there is widespread concern that the current combination of environmental stresses affecting deltas is leading to accelerated rates of relative sea-level rise. Current rates of relative sea-level rise for 46 of the world's major deltas average 6.8 mm/yr (Tessler et al. 2018) and these average changes do not capture 'hotspots' of subsidence within these deltas. This delta-specific rate is *double* that of the global trend of sea-level rise due to climate change, as measured with satellites.

The factors contributing to relative sea-level change in deltas are well understood, at least in a qualitative sense (Syvitski et al. 2009), with the change in delta surface elevation relative to sea-level being derived as the sum of the rates of natural compaction and anthropogenic subsidence, eustatic sea-level change, rate of crustal deformation due to local geodynamics, and the rate of surface aggradation (Fig. 5.1). It is important to understand that deltas naturally sink relative to sea level as a result of the natural compaction of sediments, but anthropogenic activities such as groundwater abstraction or hydrocarbon extraction often also induce accelerated subsidence (Erban et al. 2014; Brown and Nicholls 2015; Minderhoud et al. 2017). This accelerated subsidence is being compounded in many deltas by (the local expression of)

S. Hazra
School of Oceanographic Studies, Jadavpur University, Kolkata, India

M. M. Rahman
Institute of Water and Flood Management,
Bangladesh University of Engineering and Technology,
Dhaka, Bangladesh

R. J. Nicholls
School of Engineering, University of Southampton, Southampton, UK

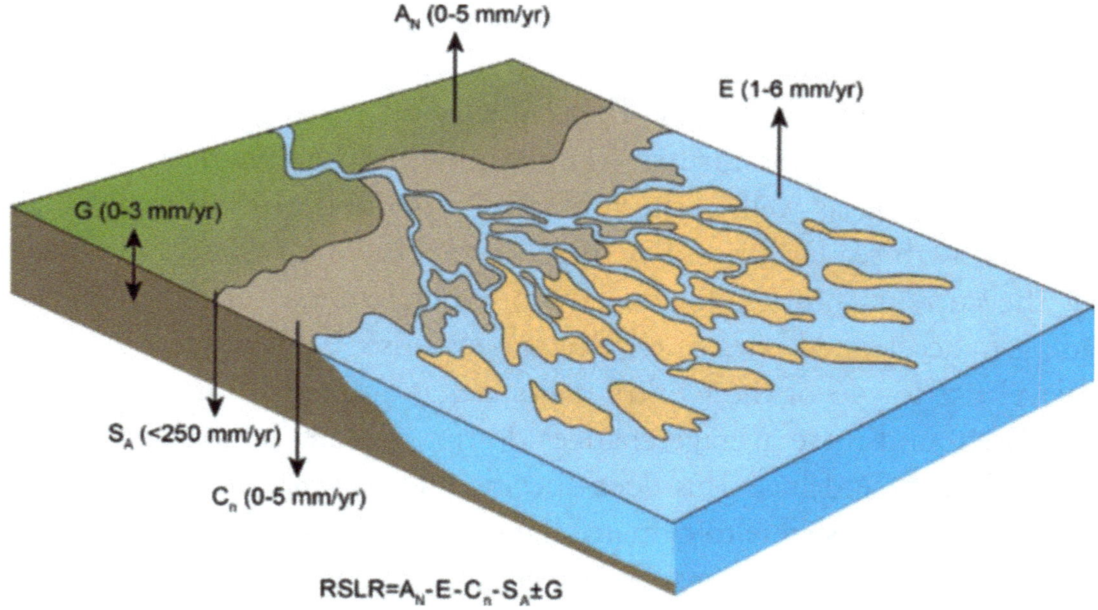

Fig. 5.1 Overview of the factors and processes contributing to relative sea-level rise within deltas. Most of the world's large deltas currently have low rates of natural sediment supply and high rates of eustatic sea-level rise (E) and often higher rates of human-induced subsidence (S_A), meaning that significant portions of the world's deltas are at risk of imminent drowning, with sediment accretion (A_N) being the only factor that can offset relative sea-level rise. Note that C_n and G denote natural compaction and crustal deformation, respectively

eustatic sea-level rise as a result of anthropogenic warming (Cazenave and Remy 2011). Thus, as human agency drives these trends, high rates of relative sea-level rise in deltas can be considered as an indicator of the Anthropocene.

Importantly, of all the drivers of relative sea-level change, the only factor that can potentially offset rising sea level, and therefore its associated risks, is sediment accretion on the delta surface. This means that it is critically important to understand the processes affecting sediment accretion—the geomorphic dynamics of deltas. In fact surface accretion rates can be understood to be affected primarily by two main factors: (i) the rate of fluvial sediment supply from the catchment upstream (Darby et al. 2016; Dunn et al. 2018; Rahman et al. 2018), and (ii) the capacity of deltas to retain that sediment (Syvitski and Kettner 2011). Yet, both of these dimensions of geomorphic dynamics of deltas are susceptible to human modification and human pressures are ubiquitous in many delta environments. This means that many of the world's major deltas may already have reached a major tipping point in which they have been shifted from a quasi-stable Holocene state to what might now be regarded as an Anthropocene (Crutzen and Stoermer 2000) state in which delta dynamics are predominantly controlled by human activity (Tessler et al. 2015).

In order to understand the drivers underlying this postulated transition from Holocene to Anthropocene deltas, alongside the potential social-ecological impacts of such a transition (Fig. 5.2), it is possible to consider the major past trends in fluvial sediment supply on major deltas. Specifically, prior studies have highlighted how fluvial sediment delivery is a critical factor in countering relative sea-level rise (Ericson et al. 2006; Syvitski et al. 2009), but that rates of sediment supply to deltas have typically declined significantly in the last few centuries as a result of catchment land use change reducing sediment yield and sediment trapping behind constructed reservoirs (Milliman and Syvitski 1992;

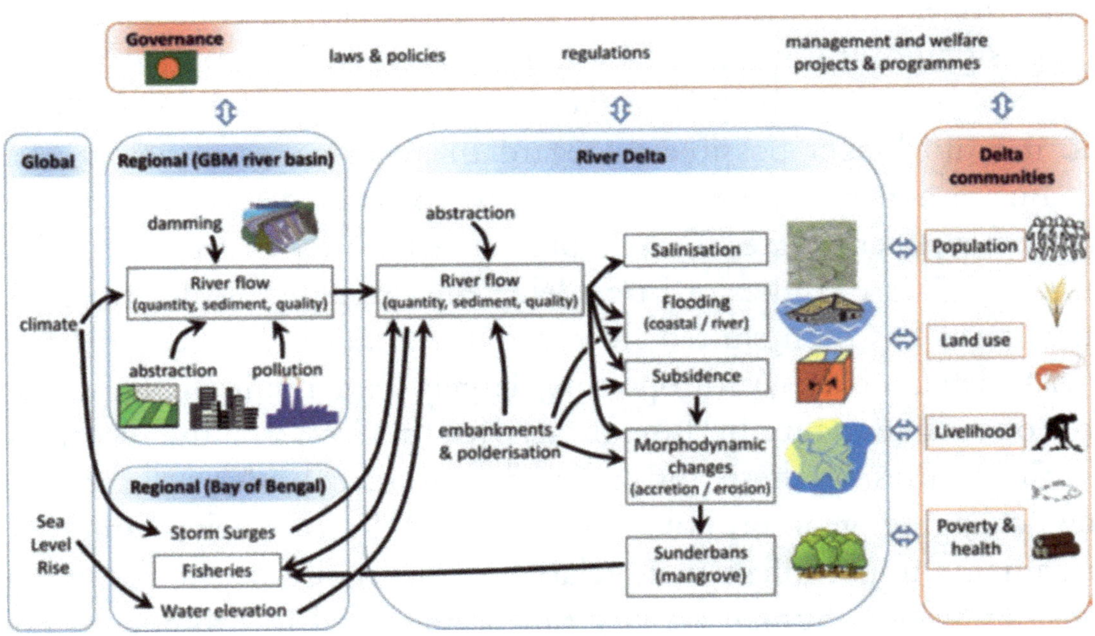

Fig. 5.2 Drivers of geomorphological change for deltas in the Anthropocene (Reprinted with permission from Nicholls et al. [2016])

Darby et al. 2016). Prior to these changes of the last few centuries, many of the world's deltas experienced *increased* rates of fluvial sediment supply associated with major land use changes driven by the onset and subsequent intensification of agriculture (Milliman et al. 1987). This points to a scenario whereby past changes in sediment flux supplied to deltas have been driven primarily by major and direct human modifications to the feeder catchments upstream of deltas, with climate-driven changes only of secondary importance (Darby et al. 2015). As discussed in this chapter, however, the radical shift in boundary conditions associated with this transition to a new Anthropocene catchment-delta system means that the effect of climate-related changes may become relatively more significant in the future.

5.2 Objectives and Overview

The aim of this chapter is to consider present and future sediment delivery and distribution in deltas, as a fundamental control on delta futures and illustrate how these processes have been, and are being, significantly impacted by human activity. It is argued that the extent of human activity is so great that, in many of the world's deltas, it is no longer possible to regard these processes as being 'natural' but rather it is more appropriate to consider deltas as existing in an Anthropocene state. The chapter is structured as follows. Below, in Sect. 5.3 case studies of three deltas are presented to illustrate how their status has been shifted from pre-disturbance Holocene deltas to at-risk deltas of the Anthropocene, largely as a result of the disruption of sediment flux by human activity. In Sect. 5.4 the 'prospects' for the futures of these and other deltas are considered, examining how changing geomorphological processes in the deltas and their catchments might affect their relative sea-level rise in the future and considering how human activity can be modified to promote a more sustainable future for at-risk deltas.

5.3 Trends of Fluvial Sediment Supply in Three Deltas

In this section the evidence base for understanding how geomorphological processes have evolved from the Holocene to the present day and beyond is presented, using the three deltas (described in Chapters 2–4) as case studies. It also examines the extent to which transitions have occurred creating an 'Anthropocene' state related to human disturbance.

The Ganges-Brahmaputra-Meghna Delta

The Ganges-Brahmaputra-Meghna (GBM) Delta (see Chapter 2, Fig. 2.1) is the second largest in the world by area, and is home to more than 100 million people (Ericson et al. 2006). The delta is fed by two major rivers, the Ganges and Brahmaputra, as well as the Meghna, with the combined catchments draining land in Bangladesh, Bhutan, China, India and Nepal. The delta itself covers the coastal region of much of Bangladesh and part of West Bengal, India, with the delta front extending for some 380 km (Allison 1998). The GBM Delta is tidally influenced, with the zone of the tidal influence extending up to 100 km inland. The many people who live in the delta live often in extreme poverty and are highly exposed to natural hazards including tropical cyclones, tidal flooding, and river and coastal erosion.

The catchments feeding the delta are located in a neo-tectonically active basin affected by block faulting and rifting which has led to frequent changes of river course and sediment supply to the delta. During the early sixteenth century, due to the easterly tilt of the basin by block faulting, the main flow of the Ganges shifted eastward and merged with the flow of Brahmaputra and then subsequently with the Meghna (Morgan and McIntire 1959) leaving the southerly flowing distributaries (e.g., Bhagirathi-Hoogly and Gorai) dry except during the monsoon period. Delta progradation, both in the subaerial and subaqueous domains continues presently only in the eastern part of the GBM Delta (i.e., the Meghna estuary) where active accretion is creating new land area (Wilson and Goodbred 2015). In contrast, the western part of the delta, especially the Hoogly or Harinbhanga estuary, has effectively been 'abandoned' (Allison et al. 2003) by the Ganges and most of the distributaries of the Ganges have now lost their upstream fresh water sources and have gradually silted up. The most recent and probably the largest direct human intervention in the delta dynamics is associated with the construction of the Farakka Barrage in 1975 (see Chapter 2), which was designed in part to redirect the dry season flow of the Ganges to the otherwise moribund Bhagirathi-Hoogly River system and the city of Kolkata. The barrage, which was built at the apex of the delta, not only initiated sedimentation upstream of the barrage, but it also considerably reduced the sediment supply downstream (Gupta et al. 2012). Field measurements show that there is now strong evidence that the combined sediment flux delivered by these two major river systems is following a declining trend (with a current rate of decline of ~10 Mt/a), such that the present fluvial sediment flux is probably about half (~500 Mt/a) of its pre-disturbance value (~1000 Mt/a) (Rahman et al. 2018).

In the last three decades, the northern part of the Bay of Bengal has recorded a sea-level rise of the order of 8 mm/yr (Pethick and Orford 2013). Indeed, two inhabited estuarine islands (Lohachara and Suparibhanga) and New Moore island in the western part of the delta have been completely lost due to a mixture of erosion and submergence (Hazra et al. 2016), reflecting low sediment supply to this part of the GBM Delta. This vulnerability is exacerbated by the GBM Delta's tendency to subsidence resulting from compaction of the deltaic deposits and anthropogenic interventions make the situation crucial to manage the delta in the coming decades. Recent studies (Pethick and Orford 2013; Brown and Nicholls 2015) show a wide range of variation of subsidence, varying from fraction of a millimetre to 45 mm/yr locally. Unable to cope with rising sea level and the reduced sediment supply (Rahman et al. 2018) the western part of the delta, in effect now abandoned by the Ganges, has suffered severe land loss to the sea over at least the last 150 years. Between 1792 and 1984, the delta lost 368 km^2 in the west while gaining 1346 km^2 in the east (i.e., around the Meghna Estuary) (Allison 1998). This wide range of variation of land change within the delta itself makes it necessary to adopt an adaptive approach to face the challenges of climate change. Thus the eastern part of the delta still maintains a supply of fluvial sediment that could enable active sediment management and restoration (see Sect. 5.4), whereas the western part is sediment starved today. Here, extensive areas of the delta will fall below mean sea level, requiring either abandonment or protective sea dykes.

The Mahanadi Delta

The Mahanadi Delta (Fig. 5.3), located in the state of Odisha on the east coast of India, is an arcuate shaped wave-dominated delta formed under a micro-tidal regime (Kumar and Bhattacharya 2003). A network of three major rivers: the Mahanadi (and its distributaries, the Devi, Daya, Bhargavi, Kushbhandra and Parchi) and the adjoining Brahmani

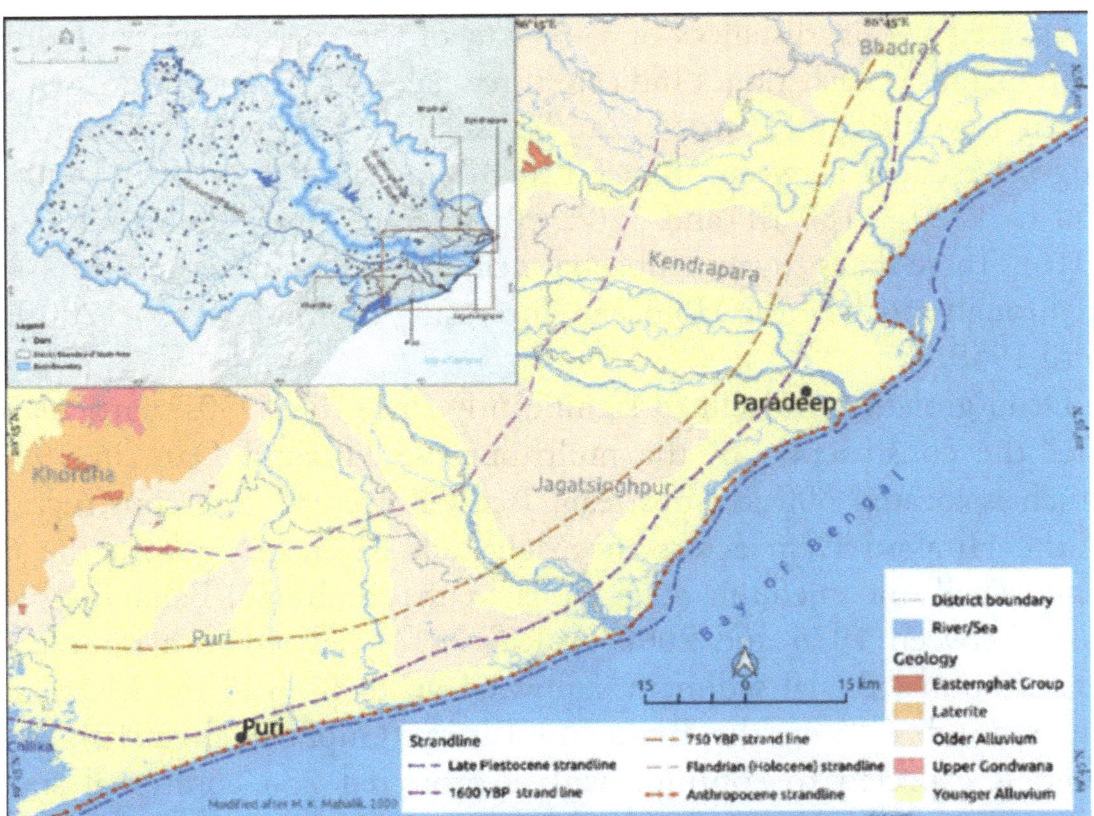

Fig. 5.3 Overview of the Mahanadi Delta and its upstream catchment illustrating the location of dams within the catchment

and Baitarini rivers contribute water and sediments to the delta. The coastline of the delta is about 200 km long, extending from the Chilika lagoon in the south to the Dhamara River in the north. The luxuriant mangrove forests of Bhitarkanika, the nesting grounds for the Olive Ridley Turtle on the spits and sandy barrier islands of Gahirmatha and the rich biodiversity of the Chilika lagoon make the delta an important biodiversity hotspot. The delta apex lies near Cuttack, approximately 90 km inland from the coast (ibid.). The delta is fed by the 851 km long Mahanadi River, which at 142,589 km^2 drains one of the largest basins of the east coast of India. The annual discharge of the Mahanadi River at the Naraj dam is 1545 m^3/s, but the delta experiences a highly variable seasonal flow regime with peak flows of up to 45,000 m^3/s generated during the southwest monsoon period. The mean annual suspended load of 27.07 Mt and bed load of 2.70 Mt is therefore supplied to the delta mainly during the monsoon months (Ray 1988; Ray and Mohanti 1989).

Located within distances of 7–35 km of the coast, a series of three shore-parallel beach ridges indicate three paleo-stand lines, from which the past evolution of the delta can be established. Moving seaward, the ridges are dated between 5880 ± 120 BP and 4250 ± 210 BP, 1590 ± 150 to 1220 ± 180 BP and 750 ± 50 BP, respectively (Somanna et al. 2016). These dates suggest subaerial delta progradation of around 4 m/yr through much of the Holocene, increasing to 9 m/yr after ~1700 BP until 1950.

Delta growth has declined significantly since the 1950s, particularly after the construction of the multi-purpose Hirakud Dam in 1957, which is located 300 km upstream (see Chapter 3). The construction of the Hirakud Dam across the Mahanadi River, along with the 60 other small and medium scale dams in the Brahmani-Baitarani river basin between 1975 and 2000 (WRIS 2014) has led to a significant loss of annual fluvial sediment supply of up to 67% in the Mahanadi and around 75% in the Brahmani Rivers (Gupta et al. 2012). This trend is expected to continue with a projected 234 additional small and medium scale dams and barrages expected to be built in the catchment over the next 40 years (Fig. 5.3). The loss of sediment supply has already been reflected in the diminished aggradation and increased erosion along a large stretch of the delta (Kumar et al. 2016), albeit with a roughly 20-year lag in the response at the shoreline following dam closure (Fig. 5.4). Specifically, historical movements of the coastline have been analysed with the End Point Rate (EPR) and Linear Regression Rate (LRR) methods and the Digital Shoreline Assessment System (DSAS) tool (Mukhopadhyay et al. 2018) of the United States Geological Survey (Fig. 5.4). These analyses indicate that, from 1990 to 2010, the shoreline receded around 90 meters near the tourist city of Puri in the south west of the delta to as much as 235 meters at Satabhaya in the north. Significant accretion (210 meters) is seen only around the port city of Paradeep where port construction and guide walls have intercepted sediment transport along the coast (Fig. 5.4). At present, 65% of the shoreline is under moderate to severe erosion, with increasing intensity from the south to north (ibid.). Forecasts of the position of the future coastline for the years 2020, 2035 and 2050 have also been undertaken, in which it was estimated that about 26 coastal

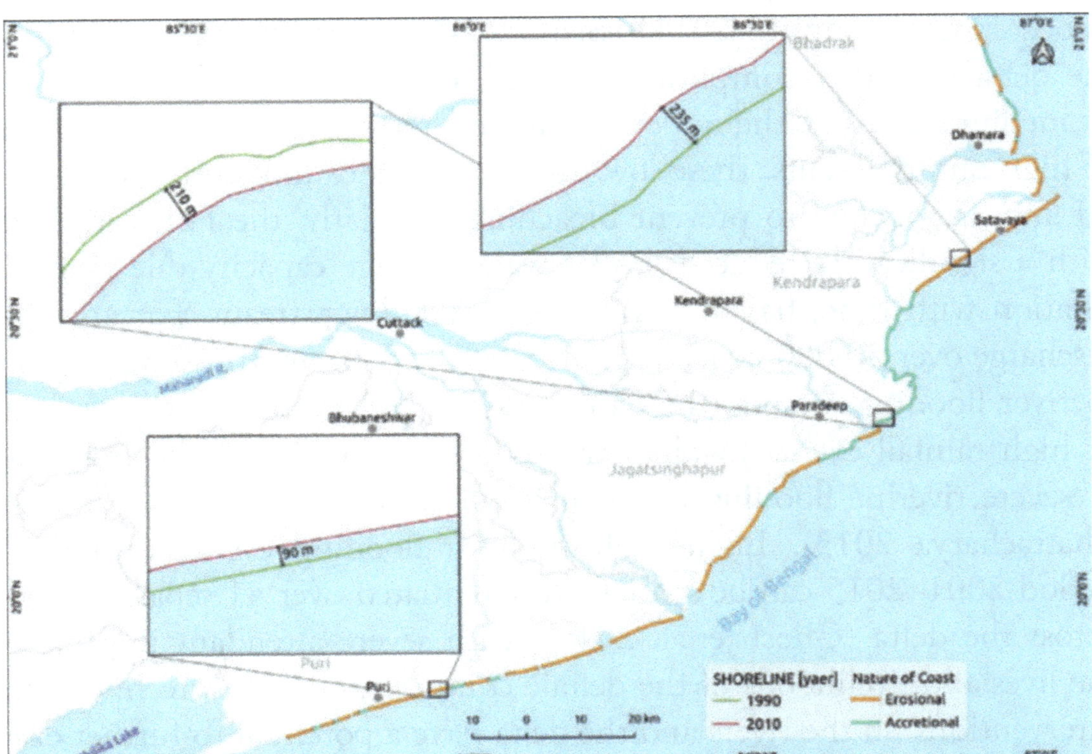

Fig. 5.4 Historical shoreline changes along the Mahanadi Delta coastal front

mouzas (small administrative blocks) out of 87 (each with an average population density of >400/km^2) would be under the threat of displacement by the year 2050 (Mukhopadhyay et al. 2018).

Changes in the progradation rate of the delta shoreline may also be driven by variations in the rate of relative sea-level change. As computed from the tide level data at Paradeep station (Source PSMSL [www.psmsl.org], 1966–2014), the relative sea-level rise seems to have accelerated in the period from 1995 to 2015 to reach a value of 8 mm/yr. At the same time, the coastal districts of the delta have witnessed a rapid growth of population and increased water withdrawal, small and medium scale irrigation projects that use both surface and ground water, deforestation of mangroves and the growth of ports, industries and urbanisation, e.g., the growth of Bhubaneswar. This means that the number of people, and the economic value of assets, exposed to flooding has increased even as relative sea-level rise is accelerating. Indeed, as with all low-lying deltas, flooding is not an unusual phenomenon on the Mahanadi Delta. Before construction of the Hirakud Dam, the delta experienced 27 devastating river floods in the 88 years between 1869 and 1957, with only nine such floods subsequently (to 2011) (Beura 2015). However, the situation in the delta is further complicated by the construction of over 200 small to medium dams in the upper catchment, mostly for irrigation. In case of high rainfall events, these low capacity dams cannot store much runoff and are opened to prevent breaching. Similarly, the Hirakud Dam, with a significantly reduced (28%) water storage capacity due to heavy siltation with time, has to pass excess water downstream. For any flow discharge over 20,000 m^3/s, large areas of the delta are thus now exposed to river flooding, despite the dams upstream. With an increasing trend of high rainfall events (Mohapatra and Mohanty 2005), the frequency of severe riverine flooding has actually increased in the delta (Jana and Bhattacharya 2013). Eight high-intensity flooding events during the period 2001–2015 can be spatially differentiated over 41 separate events across the delta (Ghosh et al. 2019), with severe attendant impacts on the lives and livelihoods of the deltaic communities. Thus, technological interventions on the rivers and the delta have a potential to further exacerbate the frequency and extent of flooding coupled with erosion and inundation along the coast in the Anthropocene.

The Volta Delta

The Volta Delta (Fig. 5.5) falls within the Keta sedimentary basin, which is one of several fault-controlled sedimentary basins in West Africa (Jørgensen and Banoeng-Yakubo 2001). The Keta basin is characterised by soft, highly compressible organic or inorganic clays overlaying fine sand, a point which is relevant to understanding contemporary controls on deltaic subsidence rates. A significant portion of the delta landscape is also characterised by the Keta Lagoon complex, the Songor lagoon, a number of creeks along the coast, extensive marsh areas, as well as extensive mangroves (see Chapter 4 and Appeaning Addo et al. 2018). The Volta River basin (390,000 km^2) has three main tributaries namely the White Volta, the Black Volta and the Oti River (Ibrahim et al. 2015) that drain a predominantly sandstone catchment that also includes a wide variety of lithologic terranes; the Volta being one of the main sources of sediment supply to the Gulf of Guinea (Goussard and Ducrocq 2014). The yearly sediment transport before the Akosombo Dam construction was about 71 million m^3, which has now reduced by about 90% following construction of the Akosombo Dam in 1965 (Boateng et al. 2012).

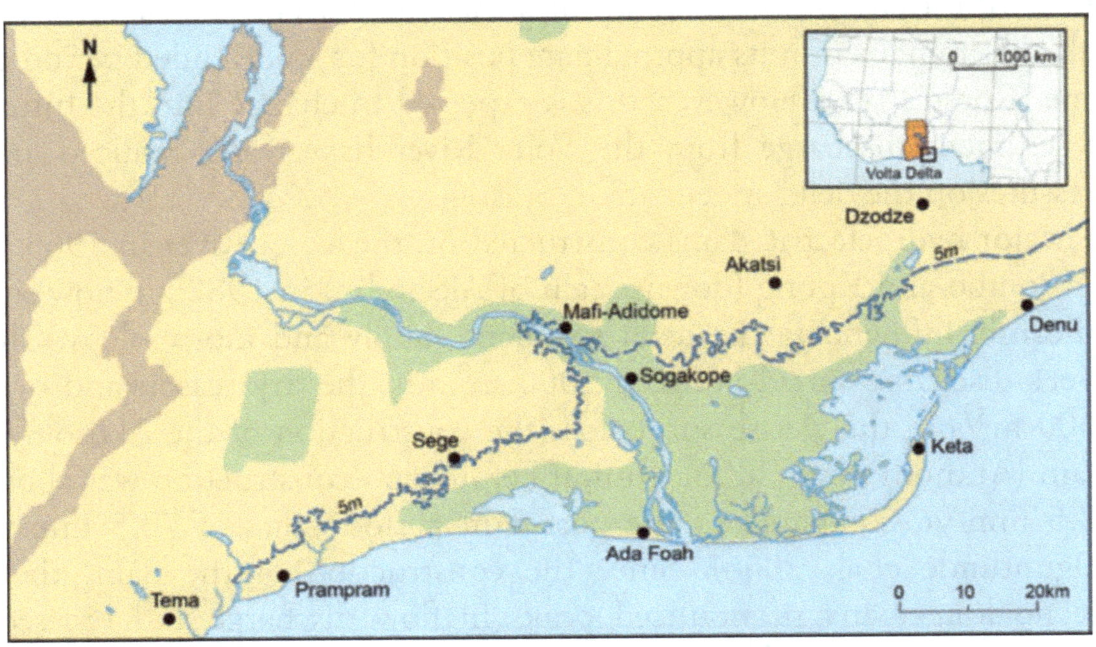

Fig. 5.5 Area of the Volta Delta in Ghana showing major settlements and the 5 m contour

The Volta River historically carried large quantities of sediment, including coarse-grained sand, to the sea and this sediment was deposited at the river mouth, forming the modern delta (Nairn et al. 1999). The surplus of sediments has built the delta lobe off the original bed-rock shoreline. The coarser sand fractions are transported by the littoral drift towards beach ridges to the east of the river mouth, while the finer mud and silt fractions, input by the Volta river, have formed a plume on the shore face east of the river mouth and on the shelf (Anthony and Blivi 1999). The Volta River has a single outlet channel to the sea at Ada with a low (1 m) tidal influence (cf. the Ganges-Brahmaputra-Meghna delta discussed in Sect. 5.3.1). The estuary is associated with a large spit that is highly dynamic. This large spit is as a result of the direct outgrowth of a natural change in the location of the mouth of the river (Anthony et al. 2016). An eastward shift, of about 12 km, in the point at which the river entered the sea has occurred since 1974 (Barry et al. 2005). The Volta Delta has features both of (river) sediment and wave domination; the sediment domination is expressed by the seaward protrusion of the delta off of the bed-rock, rather than a land-sided in-carving (Roest 2018). Carbon dating of bore-hole soil samples shows a multi-layered deltaic soil system, with peat layers dated at 5000–7000 years BP (Streif 1983). This indicates that the current perimeter of the delta has remained in its approximate position for an extended period of time (Roest 2018). However, this is expected to change into the future as sediment discharge from the Volta River has greatly reduced and pressures on the delta system are increasing.

Major hydroelectric dams constructed on the Volta River in 1965 at Akosombo and Kpong (downstream of Aksombo) in 1982 are now significantly influencing the delta sediment supply and morphology. The river's discharge varied between 1000 m^3/s in the dry season and over 6000 m^3/s in the wet season before the construction of the Akosombo Dam (Anthony et al. 2016). Runoff before dam construction was about 87.5 mm/yr, compared to the post-dam period value of 73.5 mm/yr (Oguntunde et al. 2006). Since the construction of the dam, there are no longer any pronounced peaks in flow discharge and the sediment transport is reduced to only 10% of the original transport rate (Bollen et al. 2011; Boateng et al. 2012). The reduced sediment supply to the delta system has affected the evolution of the delta (Appeaning Addo et al. 2018), shifting from a sediment-dominated towards a more wave-dominated delta (Roest 2018). As a result of the control of water discharge, flooding patterns in the delta have also changed and the annual flood-flushing of the river mouth ceased (Anthony et al. 2016). The flood plains now rely on irrigation and there is a reduced water supply for farming (Corcoran et al. 2007; Ofori-Danson et al. 2016).

Changes in the sediment flow into the delta has resulted in a significant recession of the shoreline and increased coastal flooding along the shoreline especially during energetic swell events (Allersma and Tilmans 1993). Coastal erosion in the Volta Delta was first reported in 1929, but is posited to have existed since at least the 1860s, particularly in Keta (Nairn et al. 1999). Ly (1980) estimated that the shoreline in Keta was eroding at a rate between 4 and 8 m/yr, while Bollen et al. (2011) reported that the shoreline in Totope—Ada is receding at a rate of about 6 m/yr. Kumapley (1989) estimated that a strip of coastal land about 1 km wide has been lost to erosion since 1880s. The damaging impact of erosion and flooding has resulted in the construction of engineering structures in more developed areas over the last 30 years, mainly comprising groynes and revetments, to manage the erosion and flooding in hot spot areas (Boateng 2009). Although these methods have provided localised solutions to these problems, they have also transferred the erosion and flooding problems to the down-drift coast (Angnuureng et al. 2013). Hence, in the Volta Delta, the dominant current process is erosion and sediment starvation on the open coast. Looking to the future, flooding and submergence becomes more likely as sea levels rise and subsidence continues.

5.4 Synthesis and Prospects

To conclude this overview of the geomorphological factors driving relative sea-level rise in three representative deltas, it is appropriate to consider the prospects in terms of how: (i) projected environmental changes may affect future sediment loads to each delta, and; (ii) management interventions that could be adopted to help promote sediment deposition to mitigate against sea-level rise and subsidence. Regarding the former issue, in their recent paper Dunn et al. (2018) employed a catchment model, WBMsed, to project changes in fluvial sediment delivery to each of these deltas across the twenty-first century. They also developed a series of model scenarios to represent potential future pathways of environmental change encompassing climate change, socio-economic change and reservoir construction through to the end of the twenty-first century. Specifically, the climate data used were derived from the Met Office Hadley Centre Global Environment Model version 2—Earth System (HasGEM2-ES) at 0.5 degree resolution using Representative Concentration Pathways (RCP) 2.6, 4.5, 6.0 and 8.5 from Jones et al. (2011). Reservoir data for each catchment were taken from the future dam database of Zarfl et al. (2015), which includes information on hydropower dams with over 1 MW generating capacity, both planned and under construction. Finally, the socio-economic data employed (Gross National Product [GNP] and population changes) were taken from Murakami and Yamagata (2016) to reflect the Shared Socio-economic Pathways (SSP) 1, 2 and 3. This combination of four climate and three socio-economic pathways led to a total of 12 future scenarios of change, with the reservoir construction scenarios in each case being embedded in the timelines for each future scenario (see Dunn et al. 2018).

The results indicate declining sediment loads during the remainder of this century, for the GBM and Mahanadi Deltas. Specifically, for the GBM Delta sediment loads at the end of the twenty-first century are projected to decline substantially: to 79–92 Mt/a (depending on scenario), compared to a contemporary value of ~500 Mt/a (Rahman et al. 2018). This decline is caused primarily by projected socio-economic changes in the catchment (specifically substantial increases in projected GNP which are associated with improved land management practices and lower sediment yields), with new reservoir construction being a secondary factor. At the same time, future climate change is projected to have a slight positive impact (~15% increase) on future fluvial sediment loads, partially offsetting the direct anthropogenic impacts. For the Mahanadi Delta, there is much greater variability in projected future sediment loads across the 12 investigated scenarios and the overall future relative decline in sediment load is less than for the GBM, but there is again a significant projected decline in overall fluvial sediment delivery (between ~5 and 25 Mt/a by 2100) compared to the recent past (40 Mt/a). In contrast, the Volta's trajectory is very different to that of the GBM and Mahanadi Deltas. In the case of the Volta, future changes in sediment load to the delta are negligible (changing from ~0.3 Mt/a to between 0.2 and 0.4 Mt/a depending on scenario by 2100). This reflects that the main anthropogenic impact has already occurred in one event following the construction of the Akosombo Dam in 1965, when almost all the pre-disturbance sediment supply ceased. This means that the Volta Delta already has a near zero fluvial sediment supply and there is no significant potential to restore it, assuming the continued presence of the Akosombo Dam.

In conclusion, and with regard to the role of fluvial sediment supply modulating relative sea-level rise in deltas, the global prognosis is not good, with an expectation of continued decline in sediment supply. Deltas such as the Volta and the Nile which have essentially lost their entire sediment supply due to major dams are not expected to see any positive changes. It should be noted that this prognosis does not consider climate-induced sea-level rise, which further exacerbates the situation. Moreover, it is also apparent that this prognosis is similar to the overview provided by Syvitski et al. (2009), meaning that in the ten years since that review little has changed in terms of the understanding of key anthropogenic drivers influencing relative sea-level change in deltas. Hence the prognosis for Anthropocene deltas remains one of relative loss of elevation, submergence and decline, or increasing use of engineering approaches. However, other management interventions could be successful. Ironically, if the world's deltas are seriously threatened as a result of the major human imprint on these fragile environments, then thoughtful, targeted, human agency could still offer opportunities to reverse, or at least slow, the worst impacts. For example, the Mississippi Delta has seen a failure of sediment supply and fundamental change to its hydrology and huge areas of land have been, and will continue to be, lost (Meade and Moody 2009). Nevertheless, active restoration of the Mississippi's delta is now the subject of serious investigation including major water diversions and hopes of increasing the sediment supply from the Mississippi River (e.g., Nittrouer et al. 2012). However, as losses continue it must be questioned if these efforts will reverse losses or simply slow or maybe stabilise the situation. As a contrast, the Netherlands has similar trends to other deltas, but the risk of flooding is nonetheless very low due to an intense engineering effort of both soft (on the open coast) and hard (along the rivers) measures. New governance measures have also been implemented in the form of a Delta Plan and a Delta Commission which is promoting a more adaptive and strategic approach to delta management. However, residual risk cannot be entirely eliminated and failure would be catastrophic with growing consequences over time as sea levels rise and the land sinks. Importantly, it has recently been argued that the Dutch approach may result in deltas becoming trapped in a 'dual lock-in' (Seijger et al. 2018) in which both engineering technology and institutional interests act as constraints to moving into the (arguably) more sustainable direction afforded by restoration efforts.

Looking more widely, the question is can the Dutch and/or Mississippi approaches be explored together in many deltas where there are large populations and sediment loads? Engineering can be considered as a means to promote human and economic safety and controlled flooding and sedimentation to raise land levels or build new land. What is possible needs to be assessed at the delta level and on a delta by delta basis, suggesting that the local Delta Plans (e.g., in Bangladesh for the GBM and in Viet Nam for the Mekong Delta) that are emerging need to be replicated more widely to address local circumstances and needs. These suggest the need for major human involvement in the development of populated deltas in the coming decades and longer. Tidal River Management (defined as controlled flooding and sedimentation within polders) is already practiced locally in the GBM Delta (Angamuthu et al. 2018), but the scale of application needs to increase dramatically to keep pace with relative sea-level rise. In the Mekong, flooding is now recognised as being necessary to bring sediments and nutrients to rice agriculture, which might offer a strategy for future management (Chapman and Darby 2016). More remote and relatively unpopulated deltas, such as those in the Arctic, might be left to their own devices, but even here, indirect human influence in the catchment is likely to increase and needs to be carefully considered.

References

Allersma, E., & Tilmans, W. M. K. (1993). Coastal conditions in West Africa—A review. *Ocean & Coastal Management, 19*(3), 199–240. https://doi.org/10.1016/0964-5691(93)90043-X.

Allison, M. A. (1998). Historical changes in the Ganges-Brahmaputra Delta front. *Journal of Coastal Research, 14*(4), 1269–1275.

Allison, M. A., Khan, S. R., Goodbred, S. L., & Kuehl, S. A. (2003). Stratigraphic evolution of the late Holocene Ganges-Brahmaputra lower delta plain. *Sedimentary Geology, 155*(3), 317–342. https://doi.org/10.1016/S0037-0738(02)00185-9.

Angamuthu, B., Darby, S. E., & Nicholls, R. J. (2018). Impacts of natural and human drivers on the multi-decadal morphological evolution of tidally-influenced deltas. *Proceedings of the Royal Society A: Mathematical, Physical and Engineering Science, 474,* 2219. https://doi.org/10.1098/rspa.2018.0396.

Angnuureng, B. D., Appeaning Addo, K., & Wiafe, G. (2013). Impact of sea defense structures on downdrift coasts: The case of Keta in Ghana. *Academia Journal of Environmental Science, 1*(6), 104–121. http://dx.doi.org/10.15413/ajes.2013.0102.

Anthony, E. J., Almar, R., & Aagaard, T. (2016). Recent shoreline changes in the Volta River Delta, West Africa: The roles of natural processes and human impacts. *African Journal of Aquatic Science, 41*(1), 81–87. https://doi.org/10.2989/16085914.2015.1115751.

Anthony, E. J., & Blivi, A. B. (1999). Morphosedimentary evolution of a delta-sourced, drift-aligned sand barrier–lagoon complex, western Bight of Benin. *Marine Geology, 158*(1), 161–176. https://doi.org/10.1016/S0025-3227(98)00170-4.

Anthony, E. J., Brunier, G., Besset, M., Goichot, M., Dussouillez, P., & Nguyen, V. L. (2015). Linking rapid erosion of the Mekong River Delta to human activities. *Scientific Reports, 5*, 14745. http://dx.doi.org/10.1038/srep14745.

Appeaning Addo, K., Nicholls, R. J., Codjoe, S. N. A., & Abu, M. (2018). A biophysical and socioeconomic review of the Volta Delta, Ghana. *Journal of Coastal Research, 34*(5), 1216–1226. http://dx.doi.org/10.2112/JCOASTRES-D-17-00129.1.

Barry, B., Obuobie, E., Andreini, M., Andah, W., & Pluquet, M. (2005). *The Volta River basin: Comparative study of river basin development and management* (Report) (p. 51). Colombo, Sri Lanka: International Water

Management Institute (IWMI) and Comprehensive Assessment of Water Management in Agriclture (CAWMA). http://www.iwmi.cgiar.org/assessment/files_new/research_projects/river_basin_development_and_management/VoltaRiverBasin_Boubacar.pdf. Last accessed 14 August 2018.

Beura, D. (2015). Floods in Mahanadi River, Odisha: Its causes and management. *International Journal of Engineering and Applied Sciences, 2*(2), 51–55.

Boateng, I. (2009, May 3–8). *Development of integrated shoreline management planning: A case study of Keta*. Federation of International Surveyors Working Week 2009. Eilat, Israel. https://www.fig.net/resources/proceedings/fig_proceedings/fig2009/papers/ts04e/ts04e_boateng_3463.pdf. Last accessed 14 August 2018.

Boateng, I., Bray, M., & Hooke, J. (2012). Estimating the fluvial sediment input to the coastal sediment budget: A case study of Ghana. *Geomorphology, 138*(1), 100–110. https://doi.org/10.1016/j.geomorph.2011.08.028.

Bollen, M., Trouw, K., Lerouge, F., Gruwez, V., Bolle, A., Hoffman, B., et al. (2011). Design of a coastal protection scheme for Ada at the Volta River mouth (Ghana). In *Proceedings of 32nd Conference on Coastal Engineering, 2010*. Shanghai, China: International Conference on Coastal Engineering (ICCE). http://dx.doi.org/10.9753/icce.v32.management.36.

Brown, S., & Nicholls, R. (2015). Subsidence and human influences in mega deltas: The case of the Ganges–Brahmaputra–Meghna. *Science of the Total Environment, 527,* 362–374. https://doi.org/10.1016/j.scitotenv.2015.04.124.

Cazenave, A., & Remy, F. (2011). Sea level and climate: Measurements and causes of changes. *Wiley Interdisciplinary Reviews: Climate Change, 2,* 647–662. https://doi.org/10.1002/wcc.139.

Chapman, A., & Darby, S. (2016). Evaluating sustainable adaptation strategies for vulnerable mega-deltas using system dynamics modelling: Rice agriculture in the Mekong Delta's An Giang Province, Vietnam. *Science of the Total Environment, 559,* 326–338. https://doi.org/10.1016/j.scitotenv.2016.02.162.

Corcoran, E., Ravilious, C., & Skuja, M. (2007). *Mangroves of western and central Africa*. UNEP-WCMC Biodiversity Series 26. Cambridge, UK: UNEP-Regional Seas Programme. https://www.unep-wcmc.org/resources-and-data/mangroves-of-western-and-central-africa. Last accessed 14 August 2018.

Crutzen, P. J., & Stoermer, E. F. (2000). The 'Anthropocene'. *The International Geosphere–Biosphere Programme (IGBP) Global Change Newsletter, 41,* pp. 17–18.

Darby, S. E., Dunn, F. E., Nicholls, R. J., Rahman, M., & Riddy, L. (2015). A first look at the influence of anthropogenic climate change on the future delivery of fluvial sediment to the Ganges–Brahmaputra–Meghna Delta. *Environmental Science: Processes & Impacts, 17*(9), 1587–1600. https://doi.org/10.1039/C5EM00252D.

Darby, S. E., Hackney, C. R., Leyland, J., Kummu, M., Lauri, H., Parsons, D. R., et al. (2016). Fluvial sediment supply to a mega-delta reduced by shifting tropical-cyclone activity. *Nature, 539*(7628), 276. http://dx.doi.org/10.1038/nature19809.

Dunn, F. E., Nicholls, R. J., Darby, S. E., Cohen, S., Zarfl, C., & Fekete, B. M. (2018). Projections of historical and 21st century fluvial sediment delivery to the Ganges-Brahmaputra-Meghna, Mahanadi, and Volta Deltas. *Science of the Total Environment, 642,* 105–116. https://doi.org/10.1016/j.scitotenv.2018.06.006.

Erban, L. E., Gorelick, S. M., & Zebker, H. A. (2014). Groundwater extraction, land subsidence, and sea-level rise in the Mekong Delta, Vietnam. *Environmental Research Letters, 9*(8), 084010.

Ericson, J. P., Vörösmarty, C. J., Dingman, S. L., Ward, L. G., & Meybeck, M. (2006). Effective sea-level rise and deltas: Causes of change and human dimension implications. *Global and Planetary Change, 50*(1–2), 63–82. https://doi.org/10.1016/j.gloplacha.2005.07.004.

Ghosh, A., Das, S., Ghosh, T., & Hazra, S. (2019). Risk of extreme events in delta environments: A case study of the Mahanadi Delta. *Science of the Total Environment, 664*, 713–723.

Giosan, L., Syvitksi, J. P. M., Constantinescu, S., & Day, J. (2014). Climate change: Protect the world's deltas. *Nature, 516,* 31–33. https://doi.org/10.1038/516031a.

Goussard, J. J., & Ducrocq, M. (2014). West African coastal area: Challenges and outlook. In S. Diop, J.-P. Barusseau, & C. Descamps (Eds.), *The land/ocean interactions in the coastal zone of West and Central Africa* (pp. 9–21). Heidelberg, Germany: Springer.

Gupta, H., Kao, S.-J., & Dai, M. (2012). The role of mega dams in reducing sediment fluxes: A case study of large Asian rivers. *Journal of Hydrology, 464–465,* 447–458. https://doi.org/10.1016/j.jhydrol.2012.07.038.

Hazra, S., Dey, S., & Ghosh, A. K. (2016). *Review of Odisha state adaptation policies, Mahanadi Delta. Deltas, vulnerability and climate change: Migration and adaptation (DECCMA)* (Working Paper). Southampton, UK: DECCMA Consortium. https://generic.wordpress.soton.ac.uk/deccma/resources/working-papers/. Last accessed 6 August 2018.

Ibrahim, B., Wisser, D., Barry, B., Fowe, T., & Aduna, A. (2015). Hydrological predictions for small ungauged watersheds in the Sudanian zone of the Volta basin in West Africa. *Journal of Hydrology: Regional Studies, 4,* 386–397. https://doi.org/10.1016/j.ejrh.2015.07.007.

Jana, A., & Bhattacharya, A. K. (2013). Assessment of coastal erosion vulnerability around Midnapur-Balasore Coast, Eastern India using integrated remote sensing and GIS techniques. *Journal of the Indian Society of Remote Sensing, 41*(3), 675–686. https://doi.org/10.1007/s12524-012-0251-2.

Jones, C. D., Hughes, J. K., Bellouin, N., Hardiman, S. C., Jones, G. S., Knight, J., et al. (2011). The HadGEM2-ES implementation of CMIP5 centennial simulations. *Geoscientific Model Development, 4*(3), 543–570. http://dx.doi.org/10.5194/gmd-4-543-2011.

Jørgensen, N. O., & Banoeng-Yakubo, B. K. (2001). Environmental isotopes (18O, 2H, and 87Sr/86Sr) as a tool in groundwater investigations in the Keta basin, Ghana. *Hydrogeology Journal, 9*(2), 190–201. https://doi.org/10.1007/s100400000122.

Kumapley, N. K. (1989). The geology and geotechnology of the Keta basin with particular reference to coastal protection. In W. J. M. van der Linden, S. A. P. L. Cloetingh, J. P. K. Kaasschieter, W. J. E. van de Graaff, J. Vandenberghe, & J. A. M. van der Gun (Eds.), *Coastal lowlands: Geology and geotechnology* (pp. 311–320). Dordrecht, The Netherlands: Springer.

Kumar, K. V., & Bhattacharya, A. (2003). Geological evolution of Mahanadi Delta, Orissa using high resolution satellite data. *Current Science, 85*(10), 1410–1412.

Kumar, R., Kaushik, M., Kumar, S., Ambastha, K., Sircar, I., Patnaik, P., et al. (2016). Integrating landscape dimensions in disaster risk reduction: A cluster planning approach. In F. G. Renaud, K. Sudmeier-Rieux, M. Estrella & U. Nehren (Eds.), *Ecosystem-based disaster risk reduction and adaptation in practice* (pp. 271–291). Cham, Switzerland: Springer. http://dx.doi.org/10.1007/978-3-319-43633-3_12.

Ly, C. K. (1980). The role of the Akosombo Dam on the Volta River in causing coastal erosion in central and eastern Ghana (West Africa). *Marine Geology, 37*(3), 323–332. https://doi.org/10.1016/0025-3227(80)90108-5.

Meade, R. H., & Moody, J. A. (2009). Causes for the decline of suspended-sediment discharge in the Mississippi River system, 1940–2007. *Hydrological Processes, 24*(1), 35–49. https://doi.org/10.1002/hyp.7477.

Milliman, J. D., & Syvitski, J. P. M. (1992). Geomorphic/tectonic control of sediment discharge to the ocean: The importance of small mountainous rivers. *Journal of Geology, 100,* 525–544.

Milliman, J. D., Yun-Shan, Q., Mei-E, R., & Saito, Y. (1987). Man's influence on the erosion and transport of sediment by Asian rivers: The Yellow River (Huanghe) example. *The Journal of Geology, 95*(6), 751–762. https://doi.org/10.1086/629175.

Minderhoud, P. S. J., Erkens, G., Pham, V. H., Bui, V. T., Erban, L., Kooi, H., et al. (2017). Impacts of 25 years of groundwater extraction on subsidence in the Mekong Delta, Vietnam. *Environmental Research Letters, 12*(6), 064006 http://dx.doi.org/10.1088/1748-9326/aa7146.

Mohapatra, M., & Mohanty, U. C. (2005). Some characteristics of very heavy rainfall over Orissa during summer monsoon season. *Journal of Earth System Science, 114,* 17–36.

Morgan, J. P., & McIntire, W. G. (1959). Quaternary geology of the Bengal basin, East Pakistan and India. *Geological Society of America Bulletin, 70*(3), 319–342.

Mukhopadhyay, A., Ghosh, P., Chanda, A., Ghosh, A., Ghosh, S., Das, S., et al. (2018). Threats to coastal communities of Mahanadi Delta due to imminent consequences of erosion—Present and near future. *Science of the Total Environment, 637–638*, 717–729. http://dx.doi.org/10.1016/j.scitotenv.2018.05.076.

Murakami, D., & Yamagata, Y. (2016). *Estimation of gridded population and GDP scenarios with spatially explicit statistical downscaling*. ArXiv, 1610.09041. https://arxiv.org/abs/1610.09041.

Nairn, R. B., MacIntosh, K. J., Hayes, M. O., Nai, G., Anthonio, S. L., & Valley, W. S. (1999, June 22–26). Coastal erosion at Keta Lagoon, Ghana: Large scale solution to a large scale problem. In *Proceedings of the 26th Conference on Coastal Engineering 1998*. Copenhagen, Denmark: American Society of Civil Engineers. http://dx.doi.org/10.1061/9780784404119.242.

Nicholls, R. J., Hutton, C. W., Lázár, A. N., Allan, A., Adger, W. N., Adams, H., et al. (2016). Integrated assessment of social and environmental sustainability dynamics in the Ganges-Brahmaputra-Meghna Delta, Bangladesh. *Estuarine and Coastal Shelf Science, 183*, 370–381. http://dx.doi.org/10.1016/j.ecss.2016.08.017.

Nittrouer, J. A., Best, J. L., Brantley, C., Cash, R. W., Czapiga, M., Kumar, P., et al. (2012). Mitigating land loss in coastal Louisiana by controlled diversion of Mississippi River sand. *Nature Geoscience, 5*(8), 534. http://dx.doi.org/10.1038/ngeo1525.

Ofori-Danson, B., Lawson, E. T., Ayivor, J. S., & Kanlisi, R. (2016). Sustainable livelihood adaptation in dam-affected Volta Delta, Ghana: Lessons of NGO support. *Journal of Sustainable Development, 9*(3), 1913–9071. http://dx.doi.org/10.5539/jsd.v9n3p248.

Oguntunde, P. G., Friesen, J., van de Giesen, N., & Savenije, H. H. G. (2006). Hydroclimatology of the Volta River basin in West Africa: Trends and variability from 1901 to 2002. *Physics and Chemistry of the Earth, Parts A/B/C, 31*(18), 1180–1188. https://doi.org/10.1016/j.pce.2006.02.062.

Pethick, J., & Orford, J. D. (2013). Rapid rise in effective sea-level in southwest Bangladesh: Its causes and contemporary rates. *Global and Planetary Change, 111*, 237–245. https://doi.org/10.1016/j.gloplacha.2013.09.019.

Rahman, M., Dustegir, M., Karim, R., Haque, A., Nicholls, R. J., Darby, S. E., et al. (2018). Recent sediment flux to the Ganges-Brahmaputra-Meghna Delta system. *Science of the Total Environment, 643*, 1054–1064. http://dx.doi.org/10.1016/j.scitotenv.2018.06.147.

Ray, S. B. (1988). *Sedimentological and geochemical studies on the Mahanadi River estuary, east coast of India* (Unpublished PhD thesis). Utkal University, Bhubaneswar, India.

Ray, S. B., & Mohanti, M. (1989). Sedimentary processes in the Mahanadi River estuary, east coast of India. In *Workshop on coastal processes and coastal quaternaries of Eastern India* (pp. 28–29). Calcutta, India: Geological Survey of India, Eastern Region.

Roest, L. W. M. (2018). *The coastal system of the Volta Delta, Ghana: Strategies and opportunities for development*. Delft, The Netherlands: TU Delft Delta Infrastructures and Mobility Initiative (DIMI). https://repository.tudelft.nl/islandora/object/uuid:6d859f80-e434-407e-b0bc-41f574ff8b6f. Last accessed 22 October 2018.

Seijger, C., Ellen, G. J., Janssen, S., Verheijen, E., & Erkens, G. (2018). Sinking deltas: Trapped in a dual lock-in of technology and institutions. *Prometheus*, 1–21. http://dx.doi.org/10.1080/08109028.2018.1504867.

Smajgl, A., Toan, T. Q., Nhan, D. K., Ward, J., Trung, N. H., Tri, L. Q., et al. (2015). Responding to rising sea levels in the Mekong Delta. *Nature Climate Change, 5*, 167–174. http://dx.doi.org/10.1038/nclimate2469.

Somanna, K., Somasekhara Reddy, T., & Sambasiva Rao, M. (2016). Geomorphology and evolution of the modern Mahanadi Delta using remote sensing data. *International Journal of Science and Research, 5*(2), 1329–1335.

Streif, H. (1983). Die Holozäne entwicklung und Geomorphologie der Küstenzone von Ghana. In D. Kelletat (Ed.), *Essener Symposium zur Küstenforschung, Essner Geographische Arbeiten Band 6* (pp. 1–27). Paderborn, Germany: Schöningh.

Syvitski, J. P. M., & Kettner, A. (2011). Sediment flux and the Anthropocene. *Philosophical Transactions of the Royal Society A: Mathematical, Physical and Engineering Sciences, 369*(1938), 957–975.

Syvitski, J. P. M., Kettner, A. J., Overeem, I., Hutton, E. W. H., Hannon, M. T., Brakenridge, G. R., et al. (2009). Sinking deltas due to human activities. *Nature Geoscience, 2*(10), 681–686. http://dx.doi.org/10.1038/ngeo629.

Tessler, Z. D., Vörösmarty, C. J., Grossberg, M., Gladkova, I., Aizenman, H., Syvitski, J., et al. (2015). Profiling risk and sustainability in coastal deltas of the world. *Science, 349*(6248), 638–643. http://dx.doi.org/10.1126/science.aab3574.

Tessler, Z. D., Vörösmarty, C. J., Overeem, I., & Syvitski, J. P. M. (2018). A model of water and sediment balance as determinants of relative sea level rise in contemporary and future deltas. *Geomorphology, 305,* 209–220. https://doi.org/10.1016/j.geomorph.2017.09.040.

Wilson, C. A., & Goodbred, S. L. (2015). Construction and maintenance of the Ganges-Brahmaputra-Meghna Delta: Linking process, morphology, and stratigraphy. *Annual Review of Marine Science, 7*(1), 67–88. https://doi.org/10.1146/annurev-marine-010213-135032.

WRIS. (2014). *Dams in Mahanadi.* Water Resource Information System of India. http://india-wris.nrsc.gov.in/wrpinfo/index.php?title=Dams_in_Mahanadi. Last accessed 25 October 2018.

Zarfl, C., Lumsdon, A. E., Berlekamp, J., Tydecks, L., & Tockner, K. (2015). A global boom in hydropower dam construction. *Aquatic Sciences, 77*(1), 161–170. https://doi.org/10.1007/s00027-014-0377-0.

6

Hotspots of Present and Future Risk Within Deltas: Hazards, Exposure and Vulnerability

Chris Hill, Frances Dunn, Anisul Haque, Fiifi Amoako-Johnson, Robert J. Nicholls, Pokkuluri Venkat Raju and Kwasi Appeaning Addo

6.1 Introduction

Risk, as a function of hazard, exposure, and vulnerability, is a growing characteristic of the Anthropocene concept. It is changing with evolving human assets and populations, as well as adaptation. Risk is not experienced, judged, or responded to uniformly, although often associated with levels of development and the nature of social systems (Cutter et al. 2003; Busby et al. 2014). The general tendency during the Anthropocene has been, and is likely to continue to be, low-risk tolerance in areas of high economic and investment value. Historically, this

C. Hill (✉) · F. Dunn
GeoData Institute, Geography and Environmental Science, University of Southampton, Southampton, UK
e-mail: cth@geodata.soton.ac.uk

A. Haque
Institute of Water and Flood Management, Bangladesh University of Engineering and Technology, Dhaka, Bangladesh

R. J. Nicholls et al. (eds.), *Deltas in the Anthropocene*,
https://doi.org/10.1007/978-3-030-23517-8_6

has resulted in high levels of localised protective adaptation generating protected zones which encourage further development, placing additional populations and assets at risk (e.g. Welch et al. 2017). However, risk is also fluid in nature, changing over time and in response to external influences. Mapping the spatially differentiated factors of risk such as climate variability and extremes, vulnerability of populations and natural systems to climatic stressors, and adaptive capacities provides information and understanding of changing risk as an indicator for current and future adaptation needs (Chapter 9), and locations where degradation of livelihoods may trigger migration (Chapter 7).

In addition to climate change (Chapter 1), delta regions have also been widely identified as global hotspots of vulnerability and risk due to the concentration of population and engineering interventions and the nature of these unconsolidated and dynamic coastal systems (e.g. Ericson et al. 2006; Tessler et al. 2015). Within deltas there are also local scale hotspots of risk, vulnerabilities, exposure, and hazards (e.g. Chapters 2–4). These can be hazard-specific (see Fig. 6.1), but deltas also represent environments where the interplay of physical, socio-economic, and socio-ecological systems operating over multiple spatial and temporal scales combine to produce variable levels and patterns of risk. This will inevitably continue during the Anthropocene.

F. Amoako-Johnson
University of Cape Coast, Cape Coast, Ghana

R. J. Nicholls
School of Engineering, University of Southampton, Southampton, UK

P. V. Raju
National Remote Sensing Center, Indian Space Research Organisation, Hyderabad, India

K. Appeaning Addo
Department of Marine and Fisheries Sciences, Institute for Environment and Sanitation Studies, University of Ghana, Legon-Accra, Ghana

Fig. 6.1 Hazards within deltaic areas: **a** low-lying regions close to shoreline are exposed to regular overtopping and coastal flooding (Totope, Ghana), **b** coastal and river erosion processes hazards may modify locations of risk hotspots and increase populations' vulnerability (Meghna, Bangladesh, RSC), **c** increasing subsidence and salinisation lead to changing land use and conversion from agriculture to aquaculture e.g. saline ponds (Mahanadi), **d** freshwater and brackish flooding regularly affects low-lying deltaic areas and populations, however changes in the drivers of risk can cause flood-prone locations to shift or expand (Bangladesh) (Photos: **a** Carolin Bothe-Tews; **b** Ricardo Safra de Campos; **c** Jon Lawn; **d** A. K. M Saiful Islam)

This chapter therefore explores mapping present and potential future hotspots of risk with reference to three contrasting deltas (the Ganges-Brahmaputra-Meghna [GBM], Mahanadi and Volta Deltas, Chapters 2–4) under relevant scenarios (see Kebede et al. 2018). It examines risk through its components of hazards, exposure, and vulnerabilities, reflecting both changing environmental conditions and the history of interventions and adaptations. This includes consideration of the intricately interlinked feedbacks and responses to past events and perceived risks. It also examines the spatial and temporal nature of hotspots, particularly in relation to future variation in the frequency and magnitude of hazards associated with climate change. These hotspot risk maps can help to communicate issues clearly with stakeholders and policymakers (e.g. De Sherbinin 2014; Lewis and Lenton 2015; De Sherbinin et al. 2017) and provide a basis for future research such as the sampling strategy for the household survey analysed in Chapters 7 and 9.

6.2 The Nature of Hazards in Deltas

The inherent properties of deltas such as low gradient, low elevation topography, often fertile soils (Saleque et al. 2010), and proximity to rivers and seas make them attractive for concentrations of populations and anthropogenic activities. However, these properties also increase the exposure to stresses, hazards, and multiple hazards, which compromise the biophysical and socio-economic systems of deltas (Evans 2012). These stresses and hazards occur elsewhere across the world but they are accentuated in populated deltas, where multiple processes operate and interact across spatial and temporal scales (see Chapter 1, Tables 1.1 and 1.2). Beyond the impacts on socio-economic systems in deltas, all the previously discussed hazards can affect biophysical systems and therefore ecosystem processes and services (Renaud et al. 2013). Within the context of these study deltas, the definition used is a hybrid of the morphological delta (Galloway 1975) that includes a relief characterisation (the 5 m contour) and the administrative units that intersect with the contour (Chapter 1). Although there are many potential delta definitions, this combines the morphological with socio-economic and governance components of risk. Figure 6.2 highlights the multiple natural and anthropogenic processes, and fast and slow onset events that affect delta risks, the nature of scenarios of change, and deltaic livelihoods.

The nature of hazards is often a complex of both natural and human-induced processes: for instance, salinisation levels can affect land suitability and thereby agricultural productivity and population health through contaminated ground and surface water (Syvitski 2008). In addition to being related to relative sea-level rise (RSLR), salinisation

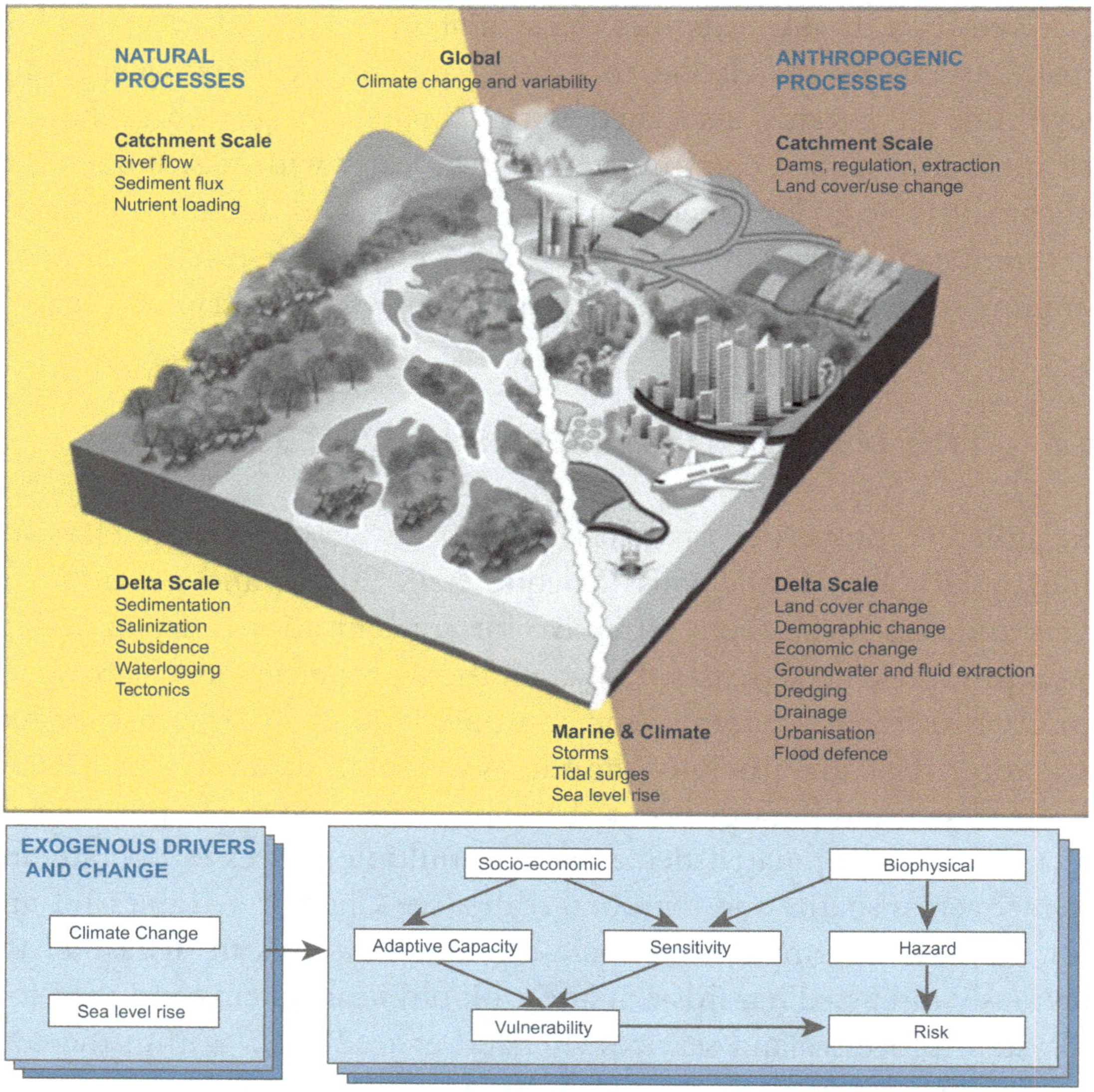

Fig. 6.2 Summary of natural atmospheric, terrestrial, marine, and anthropogenic processes in delta systems and related scenarios of risk associated with exogenous and endogenous changes

can also be caused by groundwater depletion (Erban et al. 2014), farming patterns, and agricultural practices themselves (Clarke et al. 2018). In coastal Bangladesh, salinity intrusion reflects the interplay of human and natural process, such as reductions in upstream discharge, rising sea levels and cyclonic conditions, compaction, subsidence and polderisation. Modelled scenarios of change in each of these factors show that the anthropogenic impact of intervention in upstream discharge affects the whole region, and is overlain on larger-scale exogenous factors of sea-level rise, and cyclone landfall and tracking. The resulting higher salinities, fluvio-tidal and tidal floods impact on cropland suitability and livelihoods (Kamal and Khan 2009). Equally, subsidence is due in part to natural processes of tectonic subsidence and natural compaction, but is accelerated by anthropogenic groundwater extraction, drainage of organic soils, and reduced sediment supply resultant from upstream dam construction and embanking which limits new sedimentation (Brown and Nicholls 2015; Darby et al. 2015). These interconnected factors are contributing to RSLR that increases the river and storm surge flood hazard and increases salinity and waterlogging (FAO 2015).

Climatic variations will continue and potentially become more severe with future climate change so the related hazards may also increase in frequency and severity (Janes et al. 2019). In the context of the three deltas considered in this book (Chapters 2–4), climate projections show regional increases in seasonally averaged temperature during the monsoon season (June–September) by the end of the twenty-first century, ranging from 3 to 5 °C. They also indicate an increase in average monsoon precipitation by the end of the century, ranging from 10–40% over central India. For Ghana, using data derived from Cordex (Janes et al. 2019) mean annual temperatures are projected to rise by 2.2 °C by 2050s and by 3.6 °C by 2080s across the delta area, whilst average rainfall volumes are projected to show a modest decrease. However, the projected annual potential evapotranspiration and the number of growing days are projected to decrease, with consequent impacts on rain-fed agricultural production (Fischer and Harrij 2018). Sea-level rise is also expected, reinforced by subsidence, affecting the deltas themselves (see Chapter 5).

In addition to climatic drivers, hazards in deltas are related to other external drivers such as changes in upstream catchments (Dunn et al. 2018). Economic, industrial, and land cover change in upstream catchments can influence river water quality and quantity, modifying hazards such as river flooding, hydrological drought, erosion (sediment starvation), and river nutrient levels (Whitehead et al. 2018). Changes in river nutrients can be hazardous to fishing, aquaculture, and agriculture, as well as potential human health (Syvitski 2003; Syvitski et al. 2005).

The characteristics of the deltas vary (see Table 6.1), but an improved understanding of risk and vulnerability needs to go beyond linear cause-and-effect processes, to consider the interrelated network of natural and anthropogenic processes across scales, and feedback to the hazard-producing systems themselves.

6.3 Identifying and Mapping Vulnerability and Risk Hotspots

As hazards can be multiple and spatially variable, so are social vulnerabilities (Wisner et al. 2004) and hotspots occur where concentrations of hazard and vulnerability converge to create risk. Delta communities are often dependent on climate-sensitive production systems (agriculture, fish farming, natural resources—see Chapter 8) and are subject to spatial variations in environmental stresses. Typically spatial vulnerability assessment therefore involves data integration in which geo-referenced socio-economic and biophysical data are combined with climate data to understand patterns of vulnerability.

Describing Vulnerability

Social vulnerability can be characterised by a series of factors (or domains) that are described by indicators of sensitivity (the degree to which the hazard would affect the community) and adaptive capacity (the ability of the community to cope in the short term and adjust in the longer term) to avoid the negative consequences of changes, (Amoako-Johnson and Hutton 2014; IPCC 2014). The relation can thus be expressed as:

$$\text{Vulnerability} = \text{f (Sensitivity, Adaptive Capacity)} \qquad (6.1)$$

Some indicators measure the predisposition of community vulnerability and may be directly or indirectly related to climate, but others may be largely unrelated to climate change e.g. geological hazards such as tectonics. Socio-economic factors, such as education levels or alternate livelihoods,

Table 6.1 Characteristics of the three delta systems of the GBM, Mahanadi, and Volta. Mid-century and end-century RSLR values based on RCP 8.5 for points at 2°N, 3°W (Gulf of Guinea) and 11°N, 90°E (Bay of Bengal). Based on the methodology in Palmer et al. (2018), sea levels affecting the named deltas show differences between the coastal systems from the mid-century and end-entry values, adjusted for each location

Features	GBM, Bangladesh, and India	Mahanadi, India	Volta, Ghana
Rivers/catchment area (10^3 km^2)	Ganges, Brahmaputra and Meghna (1730)	Mahanadi, Brahmani and Baitarani (141)	Black Volta, White Volta and Red Volta (398)
Study area (Chapter 1) (10^3 km^2)	51.5	12.9	5.1
Study area population (Chapter 1) (x10^6 people)	56.1	8.1	0.9
Main catchment interventions (Chapter 5)	Significant, but less affected than other two deltas to date. Farraka Barrage on the Ganges 1975	Hirakud Dam in 1957	Akosombo Dam (1961–1965) stopped all upstream influence
Current relative sea-level rise (mm/yr)	Up to 11.0	3.3	3.0
Mid-century relative sea-level rise 2041–2060	0.26 m (Bay of Bengal)	0.26 m (Bay of Bengal)	0.29 m (Gulf of Guinea)
End-century relative sea-level rise 2081–2100	0.64 m	0.64 m	0.69 m
Key current land use issues and hazards	River floods, erosion, low dry season flows, waterlogging, salinisation, surge	Floods, erosion, low dry season flows, waterlogging, salinisation, surge	Erosion (especially at Keta), floods, salinisation, drought, fire
Typical crops	Rice (main crop), wheat, jute, pulses, oilseeds, sugarcane, potatoes, vegetables, spices	Rice (main crop), wheat, jute, pulses, oilseeds, sugarcane, potatoes, vegetables, spices	Shallot, maize, cassava, tomatoes, okra, yams, rice
Typical livelihoods (see Chapter 8)	Agriculture, trade-transport, services	Agriculture, services (e.g. tourism), trade-transport	Agriculture, trade-transport, industry (e.g. salt production)
Key cities in the delta	Kolkata, Dhaka, Khulna, Chittagong	Bhubaneswar, Cuttack, Puri	None, but Accra, Ghana, and Lomé, Toga, are both adjacent

whilst independent of climate change may also be sensitive indicators of the community vulnerability and adaptability. As these are inherently hazard-specific concepts, sensitivity and adaptive capacity will change with hazard type (the factors influencing vulnerability to drought may be very different to those affected by flooding), however there is often commonality between socio-economic factors driving sensitivity such as underlying marginalisation and poverty. Critically, in terms of hotspot identification, the relationship between exposure, sensitivity and adaptive capacity vary spatially and temporally (Chen et al. 2013).

The domains and indicators of sensitivity and adaptive capacity with respect to risks are developed through stakeholder engagement at planner, land manager, community, and individual levels. The Volta Delta provides an example of the approaches used for the vulnerability mapping employing literature, stakeholder engagement, and household surveys to identify the indicators of sensitivity and adaptive capacity (Amoako-Johnson and Hutton 2014). For the Volta, ten indicators are selected to represent the sensitivity and adaptive capacity domains (Fig. 6.3) and are derived largely from enumeration level census data. The factors are inherently subjective, providing relative values of vulnerability, requiring weighting of the domain-specific relative importance; for mapping purposes, the Delphi participatory approaches were used to assign scores (Linstone and Turoff 2002). Hotspot analysis is also sensitive to the scale of the data used to identify factor variability; the highest resolution socio-economic data, based at enumeration area level, in Ghana is the 2010 census data. However, census data does not describe all the factors that may be needed to designate the domains of sensitivity and adaptive capacity. Other sources and surrogate measures, such as land cover derived from satellite information (Noor et al. 2008), may therefore be used as a measure of livelihood (i.e. dependence on agriculture and natural resources).

Methodological challenges to operating the vulnerability assessment may affect the outputs and the interpretation of the various domains of vulnerability. Data limitations, such as the lack of correspondence of data for the individual factors and multiple scales of data may affect the resolution of vulnerability analysis. Challenges include the conceptualisation of the domains, their indicators and

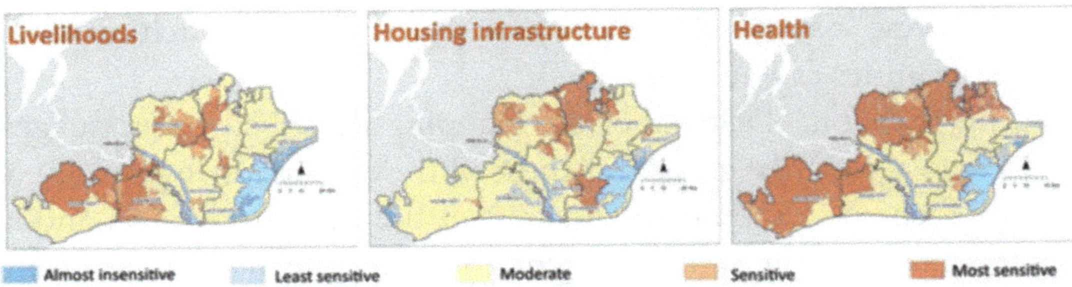

Sensitivity domains	Adaptive capacity domains
Livelihoods (dependence on agriculture production)	Access to economic alternatives and services (access to markets and services)
Housing (quality of local housing)	Human resource capacity (education, literacy and participation in the labour market)
Health (infectious disease prevalence)	Economic capacity/security (dependence on climate sensitive livelihoods)
Water and sanitation (unimproved water resources)	Enabling environment (benefiting from protection and defences)
Dependence on ecosystems services (dependence on ecosystem services)	Social capital (access to social networks)

Fig. 6.3 Geospatial variability for selected sensitivity and adaptive capacity domains and example indicators for the Volta Delta (see Amoako-Johnson et al. 2017)

measures, the degree to which data are aggregated, and the extent to which variables represented at enumeration area level are representative at the finer-scale at which hazards and stresses may be experienced by communities (Openshaw 1984). The unit of analysis issue may particularly affect communities at the coast or in riparian zones where there is exposure and susceptibility to erosional and flooding hazards in a linear zone, yet the wider sampling zone of census data include un-impacted communities. Such factors may mask significant vulnerabilities, exposure, and risk, and where vulnerability at the local level is both a predisposition to and a consequence of existing risk. This implies that the most vulnerable may have been historically exposed to risk and are likely to continue to be so in the future, in a self-reinforcing way. For example, the community on Ghana's Totope coastal barrier beach (Fig. 6.1a) are already coping and adjusting to existing hazards, are already highly exposed to storm surges and erosion hazards, and are highly vulnerable due to poverty (high sensitivity) and with limited scope to adjust livelihoods (low adaptive capacity).

However, these 'risky locations' may be the only areas which marginalised communities can access.

Within the Volta Delta, the risk mapping described above suggests that the most sensitive areas and those with least adaptive capacity are generally focused in the north and west of the delta study area, and are generally inland. Whilst delta regions include the hazards associated with their low-lying, coastal, and riverine settings (erosion, flooding), the sensitivity and adaptive capacity of communities is related to the individual environmental stresses and may reflect other factors that marginalise communities (Amoako-Johnson et al. 2017). This implies that, despite coastal hazards and sea-level rise, communities at the coast may both benefit from shoreline protection and socio-economic opportunities that overall reduce relative vulnerability.

A similar analysis of vulnerability for the Mahanadi and GBM Deltas (Ghosh 2018; Akter et al. 2019), including the future projection of the indicators (see Sect. 6.3.3) shows both the domains and socio-economic indicators for hazard-specific adaptive capacity vary based on the relevance, stakeholder evaluation, and availability of data.

Combining Vulnerability and Hazard to Map Climate Change Risk

The Fifth Assessment Report of the IPCC (2014) conceptualises risk as an objective function of the hazard, exposure to the hazard, and the socio-economic vulnerability of both assets and communities (Eq. 6.2).

$$\text{Risk} = \text{f (Hazard, Exposure, Vulnerability)} \quad (6.2)$$

Whilst the hazards are often conceived of as natural, they may be exacerbated by human-induced factors that accelerate or increase the magnitude of events or process or reduce through interventions and adaptations, for example, coastal defences, embankments, and polders may mitigate flood occurrence and flood levels (Mendelsohn et al. 2012; Haque and Nicholls 2018).

Using biophysical models under baseline and anticipated conditions assess these hazard-specific processes. For the Volta, Mahanadi, and

GBM Deltas analysed here, the external forcing of hazards is simulated by upstream hydrological and downstream ocean models driven by Met Office climate model data (Janes et al. 2019). Figure 6.4 illustrates the modelled risks for river flood, salinity, erosion, and storm surge within the Bangladesh area of the GBM Delta at the *upazila* administrative level showing both the baseline (2011) and future modelled risks to mid-century (2050). The climate change modelling in this research used downscaled Regional Climate Models RCMs Cordex Africa (0.44 degrees) and a new PRECIS South Asia downscaled data to (0.22 degrees, c. 25 km) from three selected Global Climate Models (GCMs) and based on a 'worst case' scenario using Representative Concentration Pathway (RCP) 8.5 for greenhouse gas concentrations and temperature changes developed under the Intergovernmental Panel on Climate Change Fifth Assessment Report (Janes et al. 2019). Hazard-specific risks for the future (mid-century) have been assessed by using simulations from biophysical models for storm surge, river flood, coastal erosion, riverbank erosion, and salinisation using the Delft 3D model suite, and the integrated catchment models (INCA) (Jin et al. 2018) provides the upstream boundary conditions (fluxes of freshwater water and nutrients) to secondary impact models. The sea-level rise scenarios are taken from Table 6.1.

Figure 6.4b, d, f, h illustrate the changing nature of the risk in Bangladesh for four principle hazards: river floods, salinity, erosion, and coastal storm surges. It is important to note that the changing risk is the combination of the changes in the hazards, vulnerability, and exposure components. Projections have been used for indicators where data are available or are projected e.g. female to male ratio, poverty rate. The current (baseline) flood risk in these regions of Bangladesh is mainly dominated by flood hazard; flood risk in the mid-century period increases in the northern districts of Jessore and Narail which is mainly due to increased exposure in these regions (Fig. 6.4e). The storm surge risk impact in the mid-century comes only through sea-level rise, not any change in cyclone strength which is not included. In the central region, decreased storm surge risk in the mid-century (Fig. 6.4f) is interpreted as not being due to decreased surge hazard but mainly due to decreased poverty in the region that ultimately reduces the vulnerability. Similarly, salinity risk in this region has a varied relationship to salinity hazard. Results in mid-century (right image of Fig. 6.4g) shows some decreased salinity risk in the western and central region, which is interpreted as mainly due to decreased poverty that reduces future vulnerability in the region. As described, erosion in the region is confined along the Lower Meghna and along the Tetilia system. Reduced socio-economic vulnerability has little effect on reducing future erosion risk in the region, which is evident from the risk zonation in base condition and in mid-century (Fig. 6.4h).

Combining individual risks reveals a different aspect of hotspots; as such it does not represent any specific event and may therefore have a reduced sectoral policy relevance. However, multi-risk analysis represents risk zoning comprising weighted impacts of the hazard-specific risks to reflect the aggregation of environmental stress. Distribution of multi-risk for the Bangladesh GBM Delta in baseline and in mid-century conditions (see Fig. 6.5) shows an increased total risk impact in the eastern region which is mainly due to the sustained very high storm surge risk in this region (see Fig. 6.4g). On the other hand, increased risk impact during mid-century in the western zone is due to increased risks of flood, storm surge, and salinity in this region.

Changes in agro-meteorological conditions (rainfall, temperature, evapotranspiration) also drive land use and productivity changes; National Agro Ecological Zoning (NAEZ) models simulate the agricultural yield and potential production for selected crops under changing climate and socio-economic inputs (Fischer et al. 2012; Clarke et al. 2018; Fischer and Harrij 2018). NAEZ modelling has assessed the area of suitable land relative to a baseline (1981–2010) and for two future periods (2050s and 2080s) based on the climate model inputs to provide production estimates for periods, both with and without CO_2 fertilisation (increased levels of photosynthesis resulting from increased atmospheric carbon dioxide). The baseline is for the period 1981–2010, for future climate in 2050s (ensemble mean of simulations with three climate models for the period 2041–2070) and for the 2080s (for the period 2070–2099). Taking the example of the SE Asia domain, Table 6.2 illustrates the changes in suitable area and potential production of selected crops in India (Mahanadi and GBM) and Bangladesh GBM

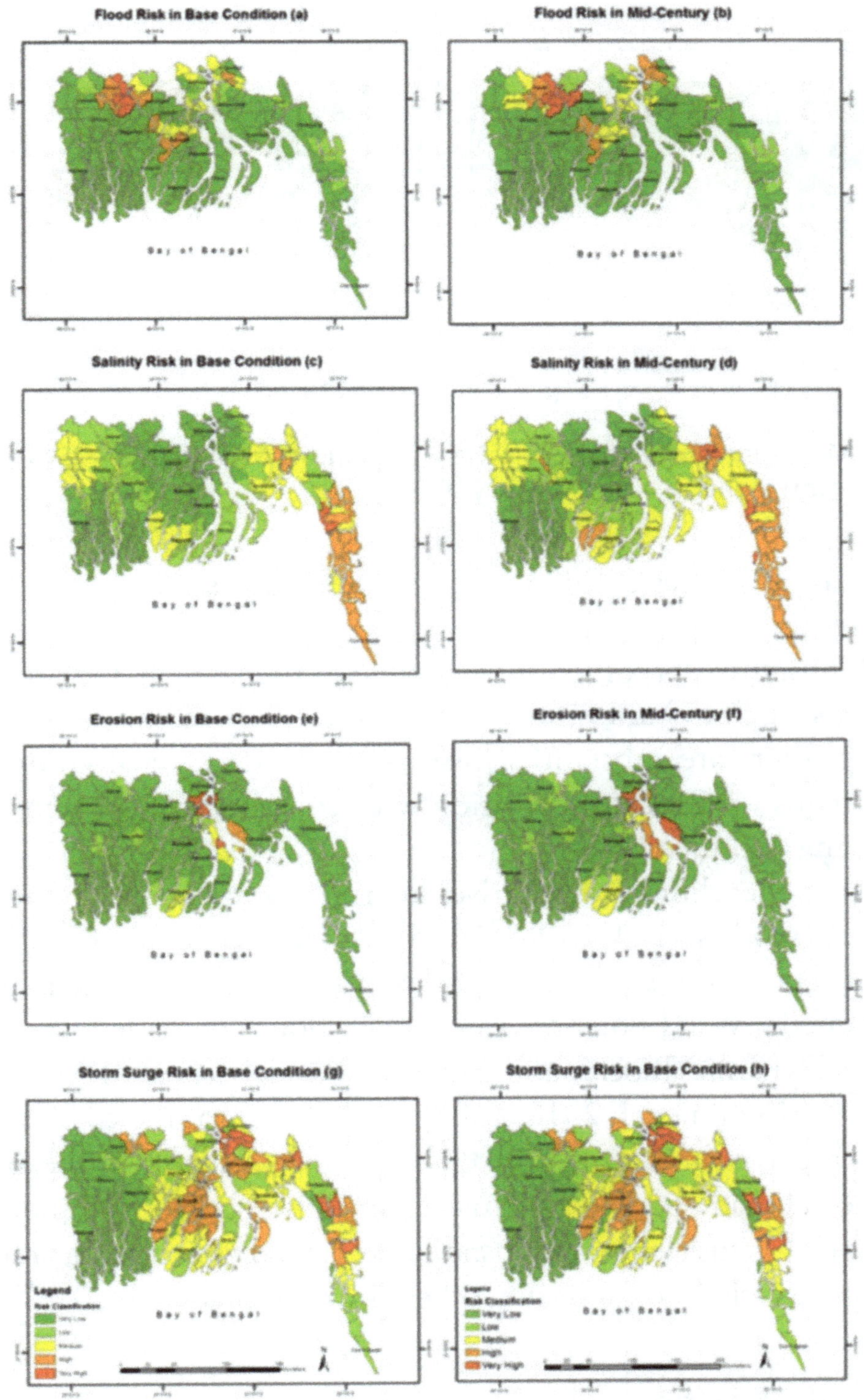

Fig. 6.4 Assessment of hazard-specific risk for the GBM Delta in Bangladesh for baseline and mid-century conditions: **a, b** river flood risk, **c, d** salinity, **e, f** erosion, and **g, h** storm surge risk. Baseline assessment is to 2001 and mid-century to 2050

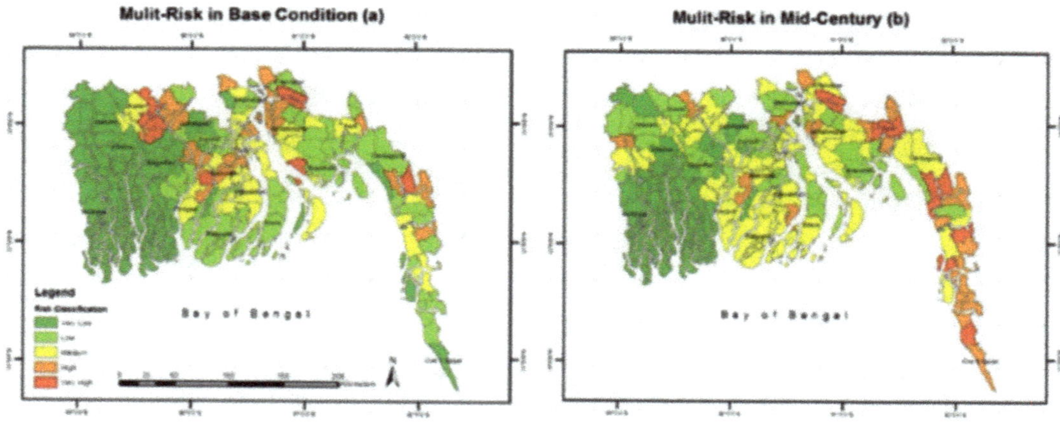

Fig. 6.5 Multi-risk assessment for the Bangladesh area of the GBM Delta (in base condition **a** and in a mid-century **b**)

Delta areas of interest. This illustrates the progressive changes in land suitability for crops, and the frequent intensification of production losses. Whilst areas become unsustainable in the later period (2080s) for some crops, such as barley, for the production of coconut, an important cash crop, there are substantial losses without CO_2 fertilisation and in contrast gains with CO_2 fertilisation with consequent divergent impacts on cropping and incomes.

Similarly, the climate data drives marine hydrodynamics and biogeochemistry models (POLCOMS-ERSEM [Kay et al. 2018]) delivering biophysical projections (sea levels and temperatures). Ecosystem and fisheries productivity and potential catch projections are made using the Size Spectrum-Dynamic Bioclimatic Envelope Model (SS-DBEM) (Cheung et al. 2009; Hossain et al. 2018; Lauria et al. 2018) for the coastal offshore zones of the Bay of Bengal and Gulf of Guinea illustrate the potential impacts of climate change and management scenarios on resources and production systems; for coastal fisheries and rural agricultural production potentials and relevance to livelihood choices (Bernier et al. 2016).

Temporal Changes in Vulnerability

As illustrated in the previous section with climate change and environmental hazards (and see Mendelsohn et al. 2012), hotspots of risk are not static. Changes to assets and socio-ecological systems, production

Table 6.2 Examples of areal extents and production potentials of rain-fed Rice, Sugarcane and Coconut for the GBM and Mahanadi Deltas in the 2050s and 2080s for baseline climate (Base) and climate scenario ensembles (ENS) with CO_2 fertilisation (ENS+) and without CO_2 fertilisation (ENS)

	Suitable extents ('000 ha)					Potential production ('000 tons)				
	Base	2050s		2080s		Base	2050s		2080s	
CO_2 fertilisation	–	ENS+	ENS	ENS+	ENS	–	ENS+	ENS	ENS+	ENS
Rain-fed rice (Kharif)										
Mahanadi	693	693	693	689	578	2656	2846	2509	2625	1746
GB India (IBD)	404	397	387	384	340	1606	1551	1345	1410	975
GBM Bangladesh	1651	1651	1650	1644	1580	7215	7502	6605	7025	5132
Sugarcane										
Mahanadi	502	227	179	76	27	2779	1198	909	379	132
GB India (IBD)	322	276	201	104	47	1785	1471	1062	510	221
GBM Bangladesh	1650	1621	1559	1345	1096	12,516	12,174	10,919	8684	6657
Coconut										
Mahanadi	695	693	481	687	379	1589	1728	1093	1776	799
GB India (IBD)	401	393	318	395	235	907	953	714	985	470
GBM Bangladesh	1659	1655	1413	1658	1304	5277	5781	4268	6093	3751

systems, population, land use change, and economic development all change over time and scale and may be even more determinants of changes in risk in the short term.

Historic analysis for the combined GBM using census data from 2001 to 2011 illustrates such changes. Figure 6.5b shows several sub-districts are socio-economically vulnerable and sub-districts like Barrackpur—I & II, Rajarhat, Daulatpur, and Raozan are socio-economically least vulnerable in both time periods. A social vulnerability gradient exists across the delta coast, where socially marginalised and vulnerable communities are found on the delta margin in both India and Bangladesh (Fig. 6.6b, see also Chapter 2). Five principal components largely determine social vulnerability in GBM Delta: (i) rural population, (ii) house ownership, (iii) agriculture dependency, (iv) lack of sanitation, and (v) existence of mud houses. Several coastal sub-districts like Koyra, Manpura, Shyamnagar, Basanti, and Morrelganj have maximum social vulnerability and have the potential to be adversely affected by environmental change, where focussed adaptation measures are immediately needed. Amongst the most vulnerable districts, Bhola, Pirojpur, Bagerhat, Shariatpur, Chandpur, and Lakshmipur show increasing trend in vulnerability ranking between 2001–2011 whilst

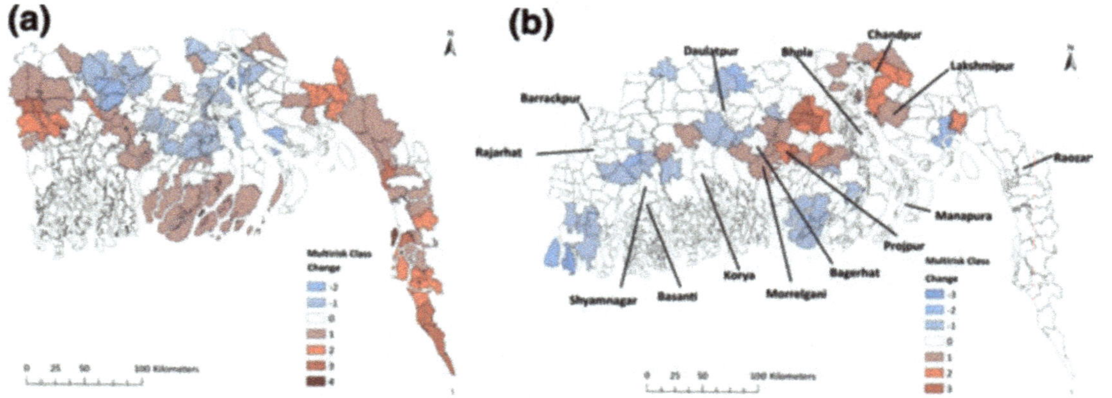

Fig. 6.6 Historical change in the location of hotspots between 2001 and 2011 based on census analysis: **a** change in multi-risk for the period in Bangladesh, **b** change in vulnerability across the combined GBM (India and Bangladesh). The figures are a relative risk rather than absolute values. Multi-risk change classes −ve scores (blue) are reducing risk/vulnerability change and +ve scores (red) indicate increased risk/vulnerability change classes

most of the other district show positive change or no change, suggesting an absence of focussed vulnerability reduction measures other than normal developmental activities.

Hypothetically, risk may be seen typically as increasing due to climate change impacts, but may be offset by adaptations and coastal/flood defences with consequent improved livelihoods and development. However, the unitary benefits gained by protection and reduced risk may collectively be lost as higher populations or assets are attracted or developed within risk zones which may increase the overall potential losses. Thus, future risk assessment requires evaluation of change scenarios from both biophysical and socio-economic contexts (Kebede et al. 2018).

6.4 Conclusion

As the Anthropocene progresses hazards are anticipated to evolve, associated with changes in long-term climate variability (such as average precipitation, temperature, and seasonal patterns), increasing number and intensity of extreme events (e.g. tropical storms, droughts), and sea-level rise/subsidence. Risks will also evolve due to socio-environmental, livelihood, assets, and infrastructure changes, reflecting the influences of economic development (e.g. urbanisation, access, irrigation), changes to production (e.g. agricultural, fisheries), changing community and household livelihoods, as well as the adaptation response to hazards and exposure. As a consequence, hotspot patterns will change over time.

Delta-level hotspot analysis offers a basis for identifying and targeting actions to support community adaptation to climate change and other stresses (Chapter 9). This is more than just mapping hazardous events. The combined geospatial analyses help to establish locations and explanations for the highest vulnerabilities and exposure and underpin effective, local actions to increase resilience. A delta-wide assessment of the multi-hazard nature and multiple indicators of vulnerability emphasises that changing risk is due to more than the obvious hazard drivers, and changes in sensitivity and adaptive capacity may counterbalance increased hazards. Within the Anthropocene there is an increasing link between the natural and man-made hazards, introducing new risks. A key conclusion based mainly on the Volta Delta analyses is the importance of non-marine climate hazards, such as drought, and the importance of non-climate vulnerabilities such as poor access to livelihood resources and development opportunities. In the Volta Delta, many risk hotspots are remote from the coastal fringe and its hazards, reflecting a wider pressure on livelihoods from long-term drought. The Volta results also suggest that access to livelihood resources, e.g. roads, is a significant influence on the vulnerability of communities. Hence, increasing access may be part of a comprehensive strategy for reducing risk. This shows that the components of hotspot risk analysis offer insights based on local community and stakeholder-defined sensitive indicators. These components are established within the context of the delta-specific environment and livelihoods systems. However, it should be noted that these relationships are associative and may not be causal.

Understanding the links and relationships between the components of vulnerability and drivers of change can have policy implications; helping to inform national and sub-national policy formulation to address the impacts of environmental changes, and the social processes and drivers that determine inequalities. Many deltas are subject to climate change policies, development strategies and delta management plans (e.g. NAPCC 2008; MoEF 2009; BDP 2100 2018). These indicate where an improved knowledge of community vulnerabilities and current adaptations can help formulate policies and support mechanisms that address the most vulnerable. As such they may also focus the opportunities and locations for adaptation, such as changing agricultural land use, crop types, and agricultural policy. Identifying potential future hotspots and potential change in hotspots can help decision-makers address the determinants of poverty and inequality and also support sustainable growth and development.

Ultimately, mapping hotspots of risk, vulnerability, and the constituents of vulnerability indicate the unequal distribution of the risk. This highlights that during the Anthropocene a wider appreciation of the nature of risk hotspots is essential in order to inform management and make the necessary adaptation choices. The trade-offs that this will entail are discussed further in Chapter 10.

References

Akter, R., Sakib, M., Najmus Sakib, M., Asik, T. Z., Maruf, M. N., Haque, A., et al. (2019). Comparative analysis of salinity intrusion in Bangladesh coast due to impacts of reduced upstream discharge, sea level rise and cyclonic condition. *Science of the Total Environment* (in press).

Amoako-Johnson, F., & Hutton, C. W. (2014). Dependence on agriculture and ecosystem services for livelihood in Northeast India and Bhutan: Vulnerability to climate change in the Tropical River Basins of the Upper Brahmaputra. *Climatic Change, 127*(1), 107–121. https://doi.org/10.1007/s10584-012-0573-7.

Amoako-Johnson, F., Quashigah, P. N. J., Hornby, D., Hill, C., Abu, M., Appeaning Addo, K., et al. (2017, July 2–5). *Uncovering climate-stressed and socially-vulnerable hotspots in the Volta Delta of Ghana: A geospatial analysis*. Deltas, Vulnerability and Climate Change: Migration and Adaptation (DECCMA) Project Presentation. Accra, Ghana: DECCMA Consortium.

BDP 2100. (2018). *Bangladesh Delta Plan 2100. Volumes 1—Strategy and 2—Investment Plan*. General Economics Division (GED), Bangladesh Planning Commission, Government of the People's Republic of Bangladesh, Dhaka, Bangladesh. https://www.bangladeshdeltaplan2100.org/. Last accessed 8 October 2018.

Bernier, Q., Sultana, P., Bell, A. R., & Ringler, C. (2016). Water management and livelihood choices in southwestern Bangladesh. *Journal of Rural Studies, 45,* 134–145. https://doi.org/10.1016/j.jrurstud.2015.12.017.

Brown, S., & Nicholls, R. (2015). Subsidence and human influences in mega deltas: the case of the Ganges–Brahmaputra–Meghna. *Science of the Total Environment, 527,* 362–374. https://doi.org/10.1016/j.scitotenv.2015.04.124.

Busby, J. W., Cook, K. H., Vizy, E. K., Smith, T. G., & Bekalo, M. (2014). *Identifying hot spots of security vulnerability associated with climate change in Africa, 124*(4), 717–731. https://doi.org/10.1007/s10584-014-1142-z.

Chen, W., Cutter, S. L., Emrich, C. T., & Shi, P. (2013). Measuring social vulnerability to natural hazards in the Yangtze River Delta region, China. *International Journal of Disaster Risk Science, 4*(4), 169–181. https://doi.org/10.1007/s13753-013-0018-6.

Cheung, W. W. L., Lam, V. W. Y., Sarmiento, J. L., Kearney, K., Watson, R., & Pauly, D. (2009). Projecting global marine biodiversity impacts under climate change scenarios. *Fish and Fisheries, 10*(3), 235–251. https://doi.org/10.1111/j.1467-2979.2008.00315.x.

Clarke, D., Lázár, A. N., Saleh, A. F. M., & Jahiruddin, M. (2018). Prospects for agriculture under climate change and soil salinisation. In R. J. Nicholls, C. W. Hutton, W. N. Adger, S. E. Hanson, M. M. Rahman & M. Salehin (Eds.), *Ecosystem services for well-being in deltas: Integrated assessment for policy analysis*. Cham, Switzerland: Springer. http://dx.doi.org/10.1007/978-3-319-71093-8_24.

Cutter, S. L., Boruff, B. J., & Shirley, W. L. (2003). Social vulnerability to environmental hazards. *Social Science Quarterly, 84*(2), 242–261. https://doi.org/10.1111/1540-6237.8402002.

Darby, S. E., Dunn, F. E., Nicholls, R. J., Rahman, M., & Riddy, L. (2015). A first look at the influence of anthropogenic climate change on the future delivery of fluvial sediment to the Ganges–Brahmaputra–Meghna delta. *Environmental Science: Processes & Impacts, 17*(9), 1587–1600. https://doi.org/10.1039/C5EM00252D.

De Sherbinin, A. (2014). Climate change hotspots mapping: What have we learned? *Climatic Change, 123*(1), 23–37. https://doi.org/10.1007/s10584-013-0900-7.

De Sherbinin, A., Apotsos, A., & Chevrier, J. (2017). Mapping the future: Policy applications of climate vulnerability mapping in West Africa. *The Geographical Journal, 183*(4), 414–425. https://doi.org/10.1111/geoj.12226.

Dunn, F. E., Nicholls, R. J., Darby, S. E., Cohen, S., Zarfl, C., & Fekete, B. M. (2018). Projections of historical and 21st century fluvial sediment delivery to the Ganges-Brahmaputra-Meghna, Mahanadi, and Volta Deltas. *Science of the Total Environment, 642,* 105–116. https://doi.org/10.1016/j.scitotenv.2018.06.006.

Erban, L. E., Gorelick, S. M., & Zebker, H. A. (2014). Groundwater extraction, land subsidence, and sea-level rise in the Mekong Delta, Vietnam. *Environmental Research Letters, 9*(8), 084010.

Ericson, J. P., Vörösmarty, C. J., Dingman, S. L., Ward, L. G., & Meybeck, M. (2006). Effective sea-level rise and deltas: Causes of change and human dimension implications. *Global and Planetary Change, 50*(1–2), 63–82. https://doi.org/10.1016/j.gloplacha.2005.07.004.

Evans, G. (2012). Deltas: The fertile dustbins of the continents. *Proceedings of the Geologists' Association, 123*(3), 397–418. https://doi.org/10.1016/j.pgeola.2011.11.001.

FAO. (2015). *Mapping exercise on water-logging in south west of Bangladesh*. Rome, Italy: Food and Agriculture Organization (FAO). https://fscluster.org/bangladesh/documents. Last accessed 28 November 2018.

Fischer, G., & Harrij, T. V. (2018). *Climate change impacts on suitability of main crops in the DECCMA study areas in Ghana and in South Asia*. Deltas, Vulnerability and Climate Change: Migration and Adaptation (DECCMA) Project Report. Southampton, UK: DECCMA Consortium.

Fischer, G., Nachtergaele, F. O., Prieler, S., Teixeira, E., Toth, G., van Velthuizen, H., et al. (2012). *Global Agro-Ecological Zones (GAEZ v3.0)—Model documentation*. Laxenburg, Austria: IIASA; Rome, Italy: FAO.

Galloway, W. E. (1975). Process framework for describing the morphologic and stratigraphic evolution of deltaic depositional systems. In M. L. Broussard (Ed.), *Deltas, models for exploration* (pp. 87–98). Houston, TX: Houston Geological Society.

Ghosh, S. (2018). *A cross-border coal power plant could put Sundarbans at risk*. The Wire.

Haque, A., & Nicholls, R. J. (2018). Floods and the Ganges-Brahmaputra-Meghna Delta. In R. J. Nicholls, C. W. Hutton, W. N. Adger, S. E. Hanson, M. M. Rahman, & M. Salehin (Eds.), *Ecosystem services for well-being in deltas: Integrated assessment for policy analysis* (pp. 147–159). Cham, Switzerland: Springer. http://dx.doi.org/10.1007/978-3-319-71093-8_8.

Hossain, M. A. R., Ahmed, M., Ojea, E., & Fernandes, J. A. (2018). Impacts and responses to environmental change in coastal livelihoods of south-west Bangladesh. *Science of the Total Environment, 637–638*, 954–970. https://doi.org/10.1016/j.scitotenv.2018.04.328.

IPCC. (2014). Climate change 2014: Impacts, adaptation, and vulnerability. Part A: Global and sectoral aspects. In C. B. Field, V. R. Barros, D. J. Dokken, K. J. Mach, M. D. Mastrandrea, T. E. Bilir, M. Chatterjee, K. L. Ebi, Y. O. Estrada, R. C. Genova, B. Girma, E. S. Kissel, A. N. Levy, S. MacCracken, P. R. Mastrandrea, & L. L. White (Eds.), *Contribution of working group II to the fifth assessment report of the intergovernmental panel on climate change* (p. 1132). Cambridge, UK and New York, NY: Cambridge University Press. http://www.ipcc.ch/report/ar5/wg2/. Last accessed 8 October 2018.

Janes, T., McGrath, F., Macadam, I., & Jones, R. (2019). High-resolution climate projections for South Asia to inform climate impacts and adaptation studies in the Ganges-Brahmaputra-Meghna and Mahanadi Deltas. *Science of the Total Environment, 650*, 1499–1520. https://doi.org/10.1016/j.scitotenv.2018.08.376.

Jin, L., Whitehead, P. G., Appeaning Addo, K., Amisigo, B., Macadam, I., Janes, T., et al. (2018). Modeling future flows of the Volta River system: Impacts of climate change and socio-economic changes. *Science of The Total Environment, 637–638*, 1069–1080.

Kamal, A. H. M., & Khan, M. A. A. (2009). Coastal and estuarine resources of Bangladesh: Management and conservation issues. *Maejo International Journal of Science and Technology, 3*(2), 313–342.

Kay, S., Caesar, J., & Janes, T. (2018). Marine dynamics and productivity in the Bay of Bengal. In R. J. Nicholls, C. W. Hutton, W. N. Adger, S. E. Hanson, M. M. Rahman, & M. Salehin (Eds.), *Ecosystem services for well-being in deltas: Integrated assessment for policy analysis* (pp. 263–275). Cham, Switzerland: Springer. http://dx.doi.org/10.1007/978-3-319-71093-8_14.

Kebede, A. S., Nicholls, R. J., Allan, A., Arto, I., Cazcarro, I., Fernandes, J. A., et al. (2018). Applying the global RCP–SSP–SPA scenario framework at sub-national scale: A multi-scale and participatory scenario approach. *Science of the Total Environment, 635*, 659–672. http://dx.doi.org/10.1016/j.scitotenv.2018.03.368.

Lauria, V., Das, I., Hazra, S., Cazcarro, I., Arto, I., Kay, S., et al. (2018). Importance of fisheries for food security across three climate change vulnerable deltas. *Science of the Total Environment, 640–641*, 1566–1577. http://dx.doi.org/10.1016/j.scitotenv.2018.06.011.

Lewis, K. H., & Lenton, T. M. (2015). Knowledge problems in climate change and security research. *Wiley Interdisciplinary Reviews: Climate Change, 6*(4), 383–399. https://doi.org/10.1002/wcc.346.

Linstone, H. A., & Turoff, M. (Eds.). (2002). *The Delphi method: Techniques and applications*, Digital edition. Boston, MA: Addison-Wesley. https://web.njit.edu/~turoff/pubs/delphibook/index.html. Last accessed 8 October 2018.

Mendelsohn, R., Emanuel, K., Chonabayashi, S., & Bakkensen, L. (2012). The impact of climate change on global tropical cyclone damage. *Nature Climate Change, 2*, 205. http://dx.doi.org/10.1038/nclimate1357.

MoEF. (2009). *Bangladesh Climate Change Strategy and Action Plan 2009*. Government of the People's Republic of Bangladesh, Dhaka, Bangladesh. Ministry of Environment and Forestry (MoEF). https://www.iucn.org/content/bangladesh-climate-change-strategy-and-action-plan-2009. Last accessed 8 October 2018.

NAPCC. (2008). *Prime Minister Council on Climate Change Report*. Ministry of Environment, Forest and Climate Change, Government of India, New Delhi, India. National Action Plan on Climate Change (NAPCC).

Noor, A. M., Alegana, V. A., Gething, P. W., Tatem, A. J., & Snow, R. W. (2008). Using remotely sensed night-time light as a proxy for poverty in Africa. *Population Health Metrics, 6*(1), 5. http://dx.doi.org/10.1186/1478-7954-6-5.

Openshaw, S. (1984). *The modifiable areal unit problem*. Concepts and Techniques in Modern Geography. Norwich, UK: Geo Books.

Palmer, M. D., Harris, G. R., & Gregory, J. M. (2018). Extending CMIP5 projections of global mean temperature change and sea level rise due to thermal expansion using a physically-based emulator. *Environmental Research Letters, 13*(8), 084003. http://dx.doi.org/10.1088/1748-9326/aad2e4.

Renaud, F. G., Syvitski, J. P. M., Sebesvari, Z., Werners, S. E., Kremer, H., Kuenzer, C., et al. (2013). Tipping from the Holocene to the Anthropocene: How threatened are major world deltas? *Current Opinion in Environmental Sustainability, 5*(6), 644–654. http://dx.doi.org/10.1016/j.cosust.2013.11.007.

Saleque, M. A., Uddin, M. K., Salam, M. A., Ismail, A. M., & Haefele, S. M. (2010). Soil characteristics of saline and non-saline deltas of Bangladesh. In C. T. Hoanh, B. W. Szuster, K. Suan-Pheng, A. M. Ismail, & A. D. Noble (Eds.), *Tropical delta and coastal zones: Food production, communities and environment at the land-water interface*. Wallingford, UK: CAB International.

Syvitski, J. P. M. (2003). Supply and flux of sediment along hydrological pathways: Research for the 21st century. *Global and Planetary Change, 39*(1), 1–11. https://doi.org/10.1016/S0921-8181(03)00008-0.

Syvitski, J. P. M. (2008). Deltas at risk. *Sustainability Science, 3*(1), 23–32. https://doi.org/10.1007/s11625-008-0043-3.

Syvitski, J. P. M., Vörösmarty, C. J., Kettner, A. J., & Green, P. (2005). Impact of humans on the flux of terrestrial sediment to the global coastal ocean. *Science, 308*(5720), 376. http://dx.doi.org/10.1126/science.1109454.

Tessler, Z., Vörösmarty, C. J., Grossberg, M., Gladkova, I., Aizenman, H., Syvitski, J., et al. (2015). Profiling risk and sustainability in coastal deltas of the world. *Science, 349*(6248), 638–643. http://dx.doi.org/10.1126/science.aab3574.

Welch, A. C., Nicholls, R. J., & Lázár, A. N. (2017). Evolving deltas: Co-evolution with engineered interventions. *Elementa Science of the Anthropocene, 5*, 49. http://dx.doi.org/10.1525/elementa.128.

Whitehead, P. G., Jin, L., Macadam, I., Janes, T., Sarkar, S., Rodda, H. J. E., et al. (2018). Modelling impacts of climate change and socio-economic change on the Ganga, Brahmaputra, Meghna, Hooghly and Mahanadi river systems in India and Bangladesh. *Science of the Total Environment, 636*, 1362–1372 http://dx.doi.org/10.1016/j.scitotenv.2018.04.362.

Wisner, B., Blaikie, P., Cannon, T., & Davis, I. (2004). *At risk: Natural hazards, people's vulnerability and disasters* (2nd ed.). Abingdon, UK: Routledge.

7

Where People Live and Move in Deltas

Ricardo Safra de Campos, Samuel Nii Ardey Codjoe, W. Neil Adger, Colette Mortreux, Sugata Hazra, Tasneem Siddiqui, Shouvik Das, D. Yaw Atiglo, Mohammad Rashed Alam Bhuiyan, Mahmudol Hasan Rocky and Mumuni Abu

7.1 Introduction

All regions of the world through the twentieth century have undergone absolute transformations in peoples' lives: they live longer, have fewer children, and are much more likely to live in towns and cities. While

R. Safra de Campos (✉) · W. N. Adger · C. Mortreux
Geography, College of Life and Environmental Sciences,
University of Exeter, Exeter, UK
e-mail: R.Safra-De-Campos@exeter.ac.uk

S. N. A. Codjoe · D. Y. Atiglo · M. Abu
Regional Institute for Population Studies, University of Ghana,
Legon-Accra, Ghana

S. Hazra · S. Das
School of Oceanographic Studies, Jadavpur University, Kolkata, India

T. Siddiqui · M. R. A. Bhuiyan · M. H. Rocky
Refugee and Migratory Movements Research Unit, University of Dhaka,
Dhaka, Bangladesh

R. J. Nicholls et al. (eds.), *Deltas in the Anthropocene*,
https://doi.org/10.1007/978-3-030-23517-8_7

these are individual decisions, it is now commonly understood that the structural drivers of such profound changes lie in economic opportunities, culturally shared expectations, access to health technologies and levels of education (Lutz et al. 2017). Regions and countries experience so-called demographic transitions of lower mortality especially infant mortality, followed by lowered total fertility rates (TFR) after a time lag, with these two combining to produce rapid population growth followed by stabilisation, population ageing and sometimes decline. The current global aggregate population of more than seven billion may stabilise over the incoming decades, but the timing of this stabilisation is not discernible in advance as it is driven by underlying drivers.

Scientific synthesis of the underlying drivers for future population shows that the outcomes that matter are on levels of fertility, mortality, migration and education, especially female education rates and quality. Analysis of the Shared Socio-economic Pathways descriptions of the future by Samir and Lutz (2017) and Abel et al. (2016) show that scenarios involving sustainability priorities, universal education and human rights result in projections of global populations around seven billion by the end of the twenty-first century. At the other extreme, future scenarios with significant and persistent poverty, population momentum, less access to education and fewer health gains, can produce global populations of greater than twelve billion by the end of the century.

A related major Anthropocene trend is in health outcomes: essentially sustained improvements in longevity and positive health across many parts of the world, despite challenging environmental conditions. McMichael (2014), for example, highlights life expectancy of more than 75 years in much of the world including in cities in countries with rapid economic growth; gains coming from sanitation, hygiene and maternal education. Limiting factors for life expectancy are currently principally non-communicable diseases such as cancers and heart and lung diseases associated with sedentary lifestyles, exacerbated by exposure to pollutants (McMichael 2014).

So why is population health, as manifest in greater life expectancy, increasing when ecosystem services are in decline and pollution exposure is increasing? This apparent paradox is explained in a number of alternative ways. On the one hand, well-being is not being reliant on the environment but principally on provisioning ecosystem services; availability of food production globally has increased substantially in the past half-century. Alternatively, health may be increasingly decoupled from environmental challenges, with urban living meaning populations are less exposed to environmental risks. Finally, as most population health research suggests, there is a significant time lag between environmental degradation and the impacts appearing in the overall health of populations (Whitmee et al. 2015). Hence McMichael (2014) and others suggest that there are looming crises and challenges to human health in the incoming decades, not least through accumulating toxic effluent in air, water and soil and through urban interfaces with climate change including exposure to heat and pollution interactions. So continued health and well-being gains from resource exploitation are not inevitable in the future, and adverse trends cannot be discounted.

The third major Anthropocene trend has involved the movement of whole populations towards urban settlement. Projections of global urban populations suggest around five billion people living in urban settlements by 2030, with an associated tripling of the global urban land area since 2000 (Seto et al. 2012). This increases the proportion of global populations living in urban areas to be greater than the current 55%, with growth concentrated in the present mega-cities, many of which are in coastal and delta areas, especially in Asia. Much of this growth occurs through migration towards urban settlements. Migration brings benefits to those moving both in terms of avoiding risks in source areas and, more importantly, providing economic, social and educational opportunities in destination areas. Individual decisions involve complex interactions between families, perceptions of opportunity and risk, and social expectations. However, at the aggregate level, the movement of many millions of people creates challenges for both rural source areas losing human capital, and for the sustainability of urbanisation processes, as illustrated in deltas (Szabo et al. 2018).

Deltas reflect these global trends and challenges. Deltas and coastal areas in general are in net population growth due to migration flows over the past half-century. They are areas where people have lived in large numbers since the advent of intensive agricultural production and have been the sites of the growth of major cities over the past century. This chapter therefore examines the demographic and mobility dimensions of development and the prospects for flourishing and sustainable futures in deltas. It does so by examining the dynamics of current and future populations in three deltas located in South Asia and West Africa—the Ganges-Brahmaputra-Meghna (GBM) which is comprised of the India Bengal and the Bangladesh sections, Mahanadi and Volta (Chapters 2–4). Given major changes in the physical and ecological dimensions of low lying coastal areas including deltas projected for the future (Brown et al. 2018), the chapter examines potential interventions to protect populations in place and to assist the movement of people away from environmental risks and harm.

7.2 Population Dynamics of Deltas

At the turn of the century around seven per cent of the world's population lived in deltas (Ericson et al. 2006). The proportion is likely to have increased since with the global phenomenon of the so-called drift to the coast. But population dynamics involve specific demographic processes such as population structures by age and sex, ongoing changes in the drivers of fertility, life expectation and mortality and migration. In addition, projections of sea-level change and rapid-onset hazards such as cyclones and flooding, exacerbate existing challenges for life and livelihoods of delta communities, as outlined in Chapter 6.

Demographic studies suggest that countries experience concomitant declining mortality and fertility levels typically associated with increasing life expectancy as they progress through the different stages of the demographic transition (Caldwell 2006; Dyson 2013). Intervening factors such as economic development and the quality of the biophysical environment influence the speed of such a transition. The majority of developing countries show this rapid demographic transition over the past 40 years associated with gradual improvement to sanitation, nutrition, access to health services and education. For example, the population structure of the coastal districts of Bangladesh (Fig. 7.1) based

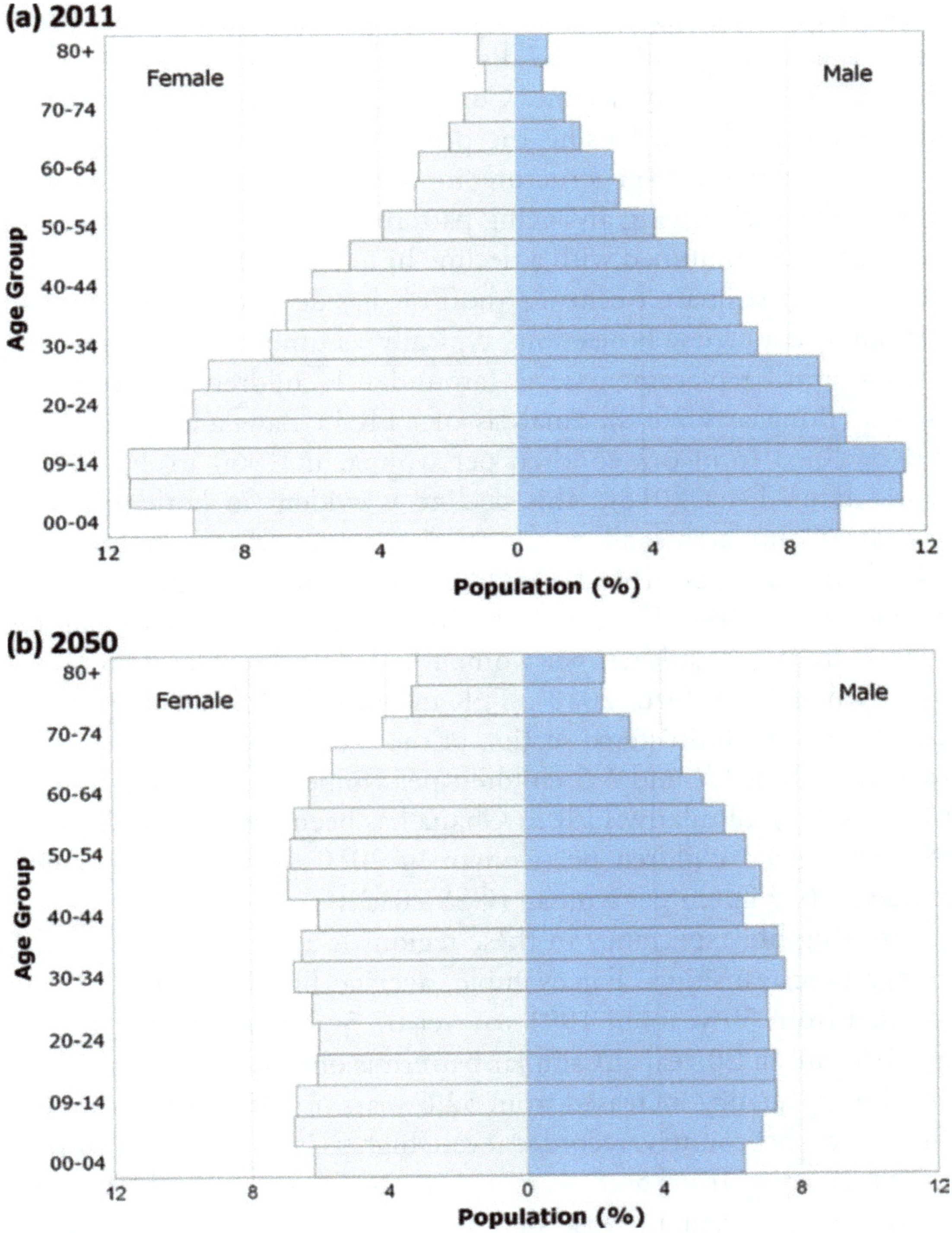

Fig. 7.1 Population pyramids for coastal Bangladesh: **a** 2011 and **b** projected to 2050 (Based on census data from the Bangladesh Bureau of Statistics [2011])

on the 2011 census (BBS 2011) is young, although it is visible that the youngest age groups (0–4) are disproportionally small, which reflects recent trends in fertility decline. Compared to the demographic projection for 2050, obtained using the projected method in the Spectrum software, the population structure in the Bangladesh portion of the GBM Delta is displaying an ageing pattern, consistent with increases in life expectancy combined with a decline in number of births.

Over the past quarter-century, the TFR has declined rapidly in many developing countries. Projections typically assume that this trend will continue until replacement level (around 2.1 children per woman) is reached (Bongaarts 2008). Analysis of TFR in Bangladesh shows that TFR declined from 6.7 children per women in 1960 to 2.1 in 2016 (World Bank Data 2018). This decline is evident in both urban and rural areas and across all administrative units, education categories, and wealth quintiles. The 2.0 children per woman in the Bangladesh portion of the GBM Delta reflects this trend. In India, national TFR declined from 5.9 children per women in 1960 to replacement level in 2016 (World Bank Data 2018). TFR in both the Mahanadi and India Bengal (Indian administered section of the GBM Delta) Deltas are even lower, recording 1.7 and 1.5 children per women, respectively (Census of India 2011). Similarly, TFR in Ghana has been continuously decreasing reaching 3.3 children per woman in 2010, while the Volta Delta recorded 3.6 children per woman (GSS 2013).

Increasing life expectancy in these regions is a key component of the demographic transition. For example, average life expectancy in India increased from 59 years in 1991 to over 67.5 years in 2011, representing 14% gain in 20 years. A similar pattern is observed in Ghana, where male life expectancy increased from 52.6 years in 1984 to 59.4 in 2010. Females in the country recorded even higher improvement, with life expectancy going from 54.8 years in 1984 to 64.4 years in 2010, resulting in a 17 per cent increase in a 24-year period. Table 7.1 includes demographic information for the three deltas (one transnational) and the national level indicators for Bangladesh, India and Ghana based on the most recent census.

Table 7.1 National and delta demographic characteristics for Bangladesh, India and Ghana

Demographic characteristic	Bangladesh		India			Ghana	
	GBM Delta, Bangladesh 2011	National 2011	India Bengal (GBM) Delta 011	Mahanadi Delta 2011	National 2011	Volta Delta 2010	National 2010
Total population (million)	40.4	150.9	18.2	8.0	1210.9	0.9	24.7
Population density (persons per km^2)	857	1023	1293	613	382	151	103
Proportion of population less than 15 years (%)	35.3	45.9	27.6	27.6	33.0	38.0	38.3
Proportion of population aged 15–64 years (%)	59.4	46.1	66.5	65.4	61.5	54.9	57.0
Proportion of population aged 65 years and above	5.3	8.0	6.0	7.0	5.5	7.1	4.7
Age dependency ratio	69	56	51	53	63	82	76
Sex ratio	97.8	100.2	95.5	96.5	94.3	87.8	95.2
Crude birth rate (per 1000 population)	18.3	19.2	11.1	16.6	21.4	28.1	25.3
Total fertility rate	2.0	2.1	1.5	1.7	2.3	3.6	3.3
Crude death rate (per 1000 population)	5.5	5.5	2.5	5.4	7.0	12.1	6.8
Infant mortality rate (per 1000 live births)	30	35	31	51	40	58	59
Under five mortality rate (per 1000 live births)	44	44	35	66	49	88	90
Proportion urban (%)	11.3	23.3	43.0	21.6	31.2	33.0	50.9
Annual population growth rate (%)	0.7	1.5	1.5	1.4	1.8	1.6	2.5

Data source Civil Registration System (CRS), 2011, Census of India; Sample Registration System (SRS), 2013, Census of India; Annual Health Survey (AHS), 2011–2012, Census of India (available only for Odisha); Tables on number of women, children ever born and child surviving, F-Series, Census of India 2011 (for indirect measures). Available from http://www.censusindia.gov.in/. BBS (2015): Population and Housing Census 2011, National Report, Volume—1, Analytical Report, Dhaka. BBS (2015): Population and Housing Census 2011, National Report, Volume—4, Socio-Economic and Demographic Report, Dhaka. BBS (2013): Sample Vital Registration System 2011. Available from http://www.bbs.dhaka.gov.bd/

Spatial Variation in Population Driven by Migration Processes

Environmental challenges result in various types of population movements driven by a wide range of spatio-temporal factors. At the same time, migration is the most challenging component of demographic projection to understand. Households employ different forms of mobility to diversify their portfolio of economic activities through access to distant labour markets in order to ensure survival or to improve their standards of living (Ellis 2000). Migration rates are highest among individuals in age group 21–30 years with secondary or higher education levels: older individuals are typically less inclined to migrate and educated people are more likely to do so (Bernard et al. 2014; Hunter et al. 2015). Gender is an important form of social differentiation that influences migration, played out in different ways across the world (Boyle and Halfacree 2002). Migration intentions and propensity are also manifest in roots in the form of place attachment to where people live (place attachment—Adams 2016) and the perceived benefits from migration (de Jong and Fawcett 1981).

Unequal spatial development, socio-economic transformations and environmental stress are key factors driving ongoing population movement in deltas worldwide (Seto 2011). Deltas are often described as popular destinations for migrants because of diverse socio-economic opportunities associated with fertile soils, freshwater resources and rich biodiversity conducive to rapid agricultural development (Renaud et al. 2013). Such characteristics combine to act as a pull factor attracting migrants to areas that are already densely populated. This rapid consolidation of population has transformed the major global deltas from agrarian economies into industrial and service-driven cities which are projected to continue to act as hubs of attraction to new populations therefore increasing the pressure on delivery of services (Nicholls et al. 2018).

Despite current high rates of urbanisation and the range of factors described above acting as pull factors, net migration rates for the GBM, Mahanadi and Volta Deltas, based on the two most recent censuses for Bangladesh, India and Ghana, reveal contrasting evidence of in-migration as shown in Fig. 7.2. The Bangladeshi GBM Delta is a net sender of migrants to other areas of the country particularly to Dhaka and Chattogram

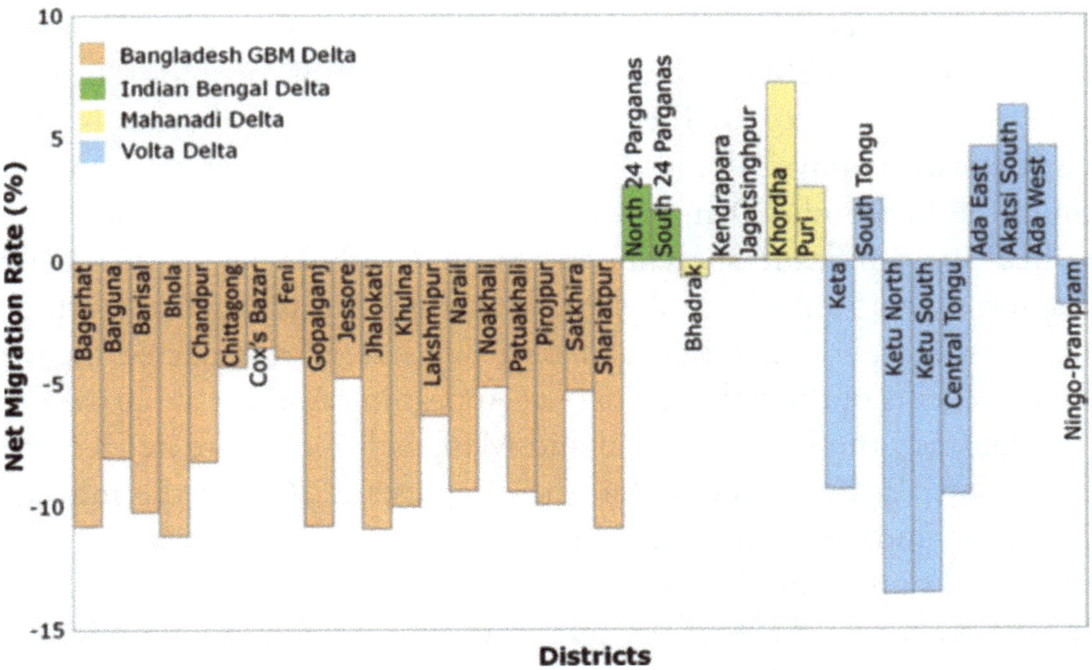

Fig. 7.2 Net migration rate by delta district in Bangladesh, Ghana and India

(city formerly known as Chittagong) and there is more out-migration of males compared to females. This is similar to the trend recorded in the Mekong Delta, in Vietnam, which has recorded high out-migration rates (Szabo et al. 2016). In addition, all 19 districts located in the Bangladeshi GBM Delta recorded negative net migration. For example, Bhola, an isolated island district, recorded the highest out-migration between 2001 and 2011 (11.2% of total population); Cox's Bazar recorded the lowest net out-migration for the same period (3.6% of total population). This net migration is explained in large part by internal migration to the two major urban centres of the country, Dhaka and Chattogram, where employment and access to services are offered in higher proportion compared to rural areas. Dhaka had net annual migration arrivals between 2000 and 2010 of 300–400,000 (te Lintelo et al. 2018), while Chattogram had an aggregate population of 3.3 million in 2001 and four million in 2011 with an annual growth rate of 3.6% between 1990 and 2011 (Mia et al. 2015). The majority of new arrivals to Chattogram are associated with economic opportunities brought about by rapid industrialisation such as garment manufacture (which is responsible for two-thirds of the growth in employment in Chattogram [Mia et al. 2015]).

The Volta Delta also has a negative net migration of about 41,000 people over the period 2000–2010, representing 4.8% of the mid-year population of the area in 2010. The negative net migration in the area is partly as a result of biophysical and development factors such as the construction of the Akosombo Dam in 1964 and the Akuse Dam in 1982, which resulted in significant shoreline recession (Anthony et al. 2016) (see Chapters 4 and 5). Additionally, environmental hazards such as land subsidence, sea-level rise and saltwater intrusion have also impacted the lives of the residents of the Volta Delta (e.g. Appeaning Addo et al. 2018).

In contrast, the Indian Bengal and Mahanadi Deltas are net receivers of migrants. In the former delta, these represent 2.6% of the mid-year population and are evident in the high proportion of urbanisation (43% of the total population live in urban areas as shown in Table 7.1). In the Mahanadi Delta, 2.51% of the total mid-year population are migrants. Figures 7.3 and 7.4 show the direction of migrant flows from both the Indian Bengal and the Mahanadi Deltas. The widths of arrows represent the number of individuals who reported relocating to a different state within the country based on data from a cross-sectional household survey conducted for this research in India in 2016. This reflects that societies tend to become more mobile as the level of development of countries improve; a trend in line with the mobility transition theory postulated by Zelinsky (1971) and examined by, for example, de Haas (2010) and Skeldon (2014). It is relevant to mention that the population movement examined in this section capture long term moves including permanent, seasonal and circular mobility. Previous research shows that empirical evidence on other forms of temporary moves that form the broad spectrum of the mobility continuum such as daily, weekly and occasional movements is more limited (Safra de Campos et al. 2016) and therefore were not included in this research.

Growth of Urban Centres and Impacts of Migration on Rural–Urban Linkages

Associated with population dynamics and increased mobility is the growth of cities, particularly mega-cities in deltas, with the majority of the global population at present living in urban areas

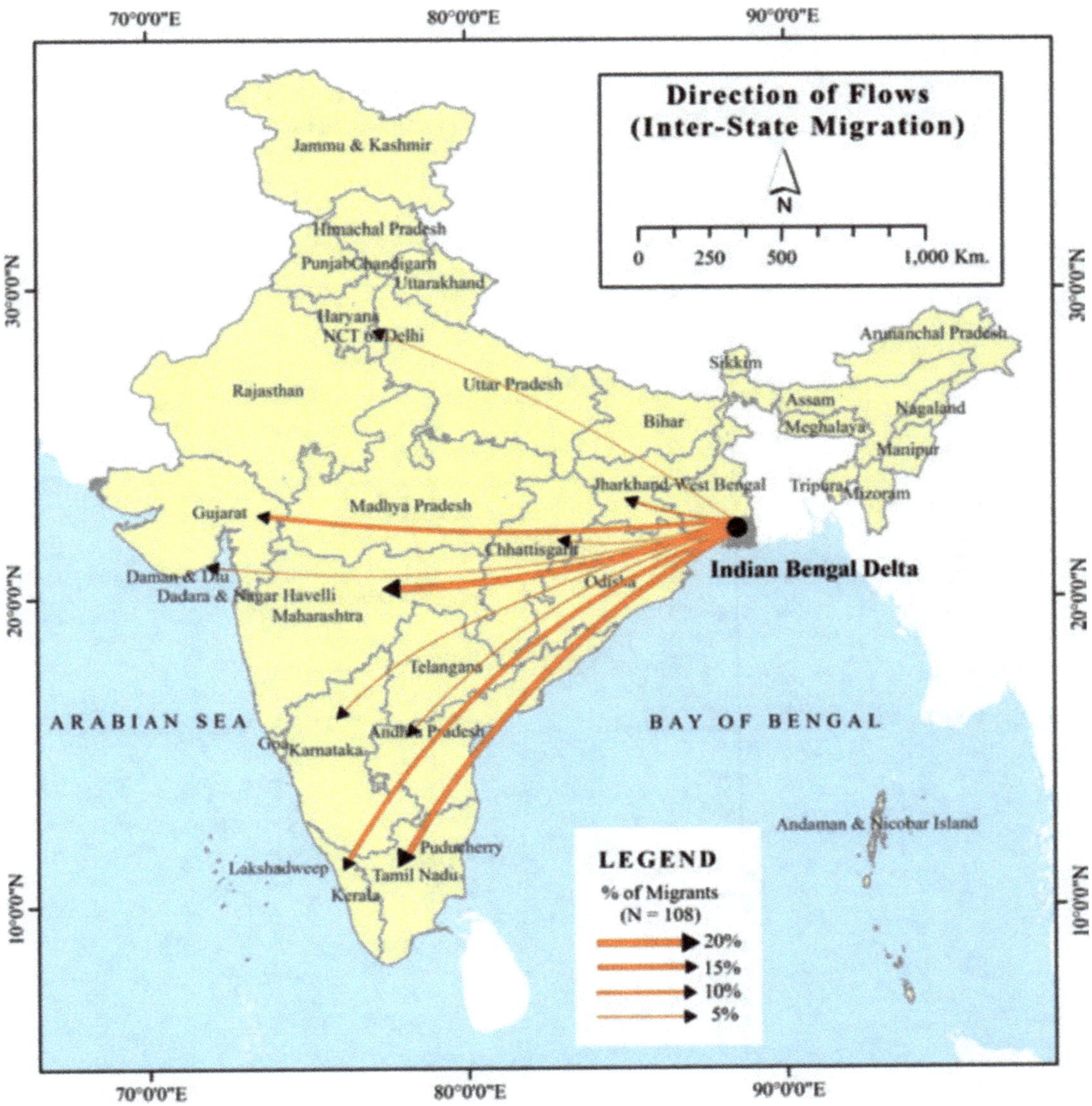

Fig. 7.3 Direction of interstate migration from the Indian section of the GBM Delta, India ($N = 108$)

(World Bank Group 2014). Seto (2011) suggests that cities in deltas face two main challenges. First, future urban growth will be more significant in Asia and Africa resulting in additional pressure on existing infrastructure and provision of services. Second, climate change is predicted to increase the frequency and magnitude of extreme events leaving low-lying coastal zones and deltas exposed to the impacts of storm surges, sea-level rise, flooding and coastal erosion (Brown et al. 2018; Nicholls et al. 2018).

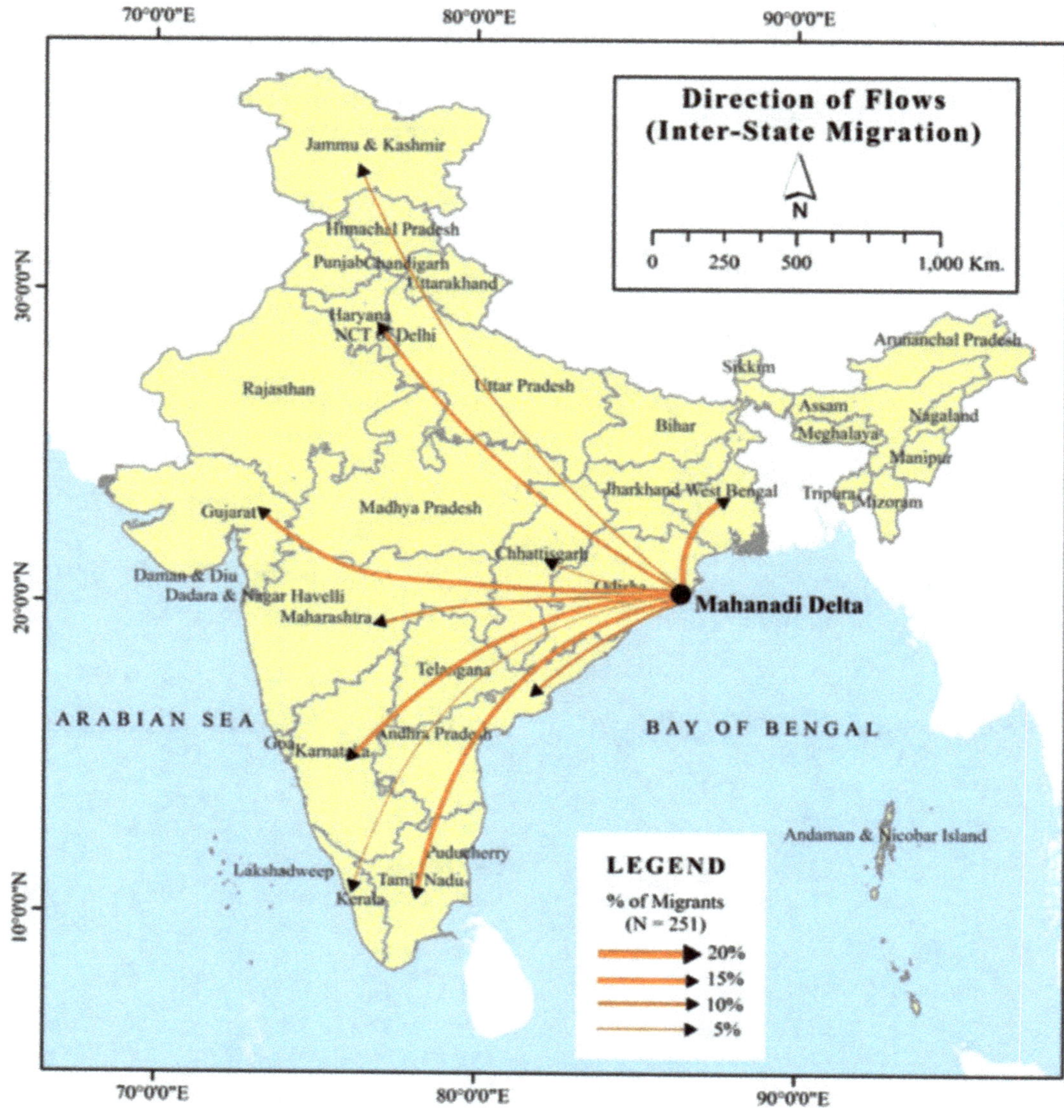

Fig. 7.4 Direction of interstate migration from the Mahanadi Delta, India ($N = 251$)

At the same time, deltas are undergoing substantial land-use changes, for example, ports and port-related industries, which have dominated economic and spatial development of many of coastal areas (Dammers et al. 2014). The introduction of alternative ecosystem-based livelihoods including intensive aquaculture has brought about impacts associated with land tenure and livelihood displacement resulting in income loss, food insecurity, rural unemployment and forced migration (Hossain et al. 2013; Amoako-Johnson et al. 2016).

In addition to the socio-economic and environmental challenges, there are other dimensions associated with rapid urbanisation. Population movements are often associated with unequal development and distribution of resources (Tacoli and Mabala 2010). However, the flow of remittances between urban and rural areas is an important and stable economic resource that supports communities in rural areas. The importance of remittances for the domestic economy of households in deltas was captured in the household survey conducted in Bangladesh, Ghana and India. Across these three deltas, 66% of respondents reported receiving remittances. The vast majority of respondents used this income mainly to help pay for food, health, education, debt repayment and household appliances. This supports widespread evidence that remittances sent by migrants have the potential to enable rural households to overcome credit and risk constraints by spatial diversification of labour and income (Stark 1991). Moreover, if invested in modern agricultural technologies, tools and livestock, subsistence farmers can increase their productivity and complete the transition from familial to commercial production, which is instrumental in the diversification of rural economies (Webber and Barnett 2010).

In addition to financial benefits derived from remittance income, migration also circulates new ideas, knowledge, skills and technologies from destination to origin areas (Levitt and Lamba-Nieves 2011). This rural–urban linkage through social remittances was also expressed in the same survey; 75% of households reported benefits from new ideas and knowledge to build adaptive capacities at the origin.

7.3 Environmental Stress: Trigger for Migration?

Deltas and low-lying coastal areas are at risk from sea-level rise and storm surges which may result in submergence of seafront settlements and increased flooding of coastal land, as well as saltwater intrusion of surface waters and groundwater (Nicholls and Cazenave 2010; Brown et al. 2018). Changes in the intensity and frequency of cyclones and persistent variance in pluviometry are also a likely consequence of climate change. Flooding in low-lying densely populated coastal areas is predominantly seasonal and usually short-lived, yet it can have significant impacts on vulnerable sections of society. For example, poor households living in flood prone areas might become displaced and forced to move temporarily seeking access to frontline services and alternative forms of livelihood. Furthermore, there might be instances where impacted communities must be relocated permanently. Due to the characteristics described above, it is tempting to suggest that existing or future environmental factors are influential in driving individual decisions on migration, more so in already vulnerable locations in low-lying coastal areas. The cross-sectional survey of 5479 households across the delta locations in India, Ghana and Bangladesh revealed that more than 30% of these representative households reported at least one migrant. The survey collected data on the motivation for migration by one or more household members: eliciting all relevant motivations for migration and ranking them to ascertain the main driver reported by household heads. Figure 7.5 shows that only 2.8% of respondents perceived the main reason behind the decision to move to have been associated with environmental stress. The majority of respondents perceived economic and social factors to have accounted for the migration of their household members.

The results reveal that few people, even in places significantly exposed to environmental hazards and climate variability, self-report as environmental migrants. Yet environmental risks may still play a significant role in migration decisions at the household level (Abu et al. 2014). In the aforementioned survey, households in coastal Ghana, India and Bangladesh also reported significant exposure to environmental risks and perceived economic insecurity associated with environmental hazards. One-third of all respondents perceived that there was an increased exposure to hazards including cyclone, drought, erosion, flooding, salinity and storm surge over the previous five years. Over one-third of respondents (37.5%) also reported the environment to be more hazardous and extreme events to be more frequent. Similarly, between

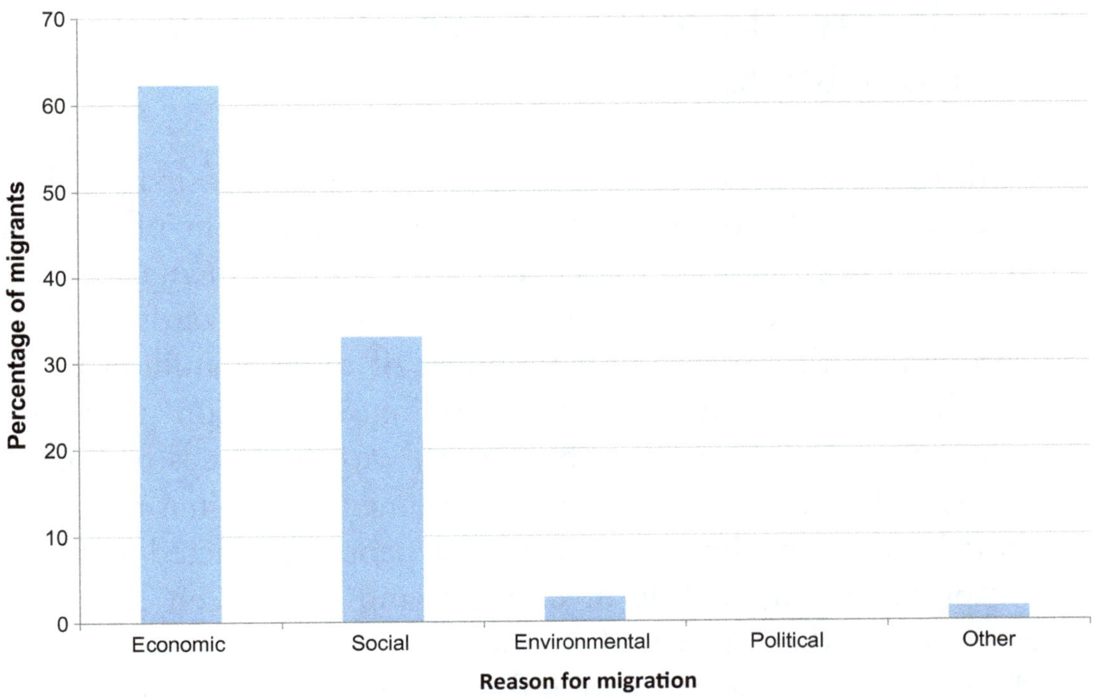

Fig. 7.5 Main reason for migration across the three deltas ($N = 2310$): economic (employment and debt); social (education, marriage, health, family and housing problems); environmental (loss of income for one/multiple seasons, environmental degradation and extreme events); political (social/political problems)

40 and 80% of the respondents associated environmental factors with more insecure livelihoods. These perceptions of underlying environmental degradation and insecurity suggest a correlation with migration of household members.

Analysis also recognised the diversity of environmental stressors, categorising these as either rapid- or slow-onset, each potentially affecting migration in different forms (Warner 2010) with links being made between migration and thresholds associated with access to various forms of capital assets, and changes in land use and livelihood (McLeman 2018b). This shows that the role played by mobility as a response to climate events depends heavily on the duration, intensity and nature of the stimulus, as well as the composition and assets of households, their previous experience, the networks to which they belong, and the adaptation responses set in motion (see Chapter 9).

7.4 Migrate, Relocate or Remain: Policy and Interventions in Deltas

The Anthropocene trend of increased human mobility may well be intensified by the effects of climate change, displacement often being one of the consequences associated with environmental change. Displacement due to climate and environmental hazards is common throughout the world, with estimates of over 18 million people impacted by weather-related events, including 8.6 million people displaced by floods and 7.5 million by storms, with hundreds of millions more at risk (iDMC 2018). Resulting movements are often temporary and short-lived, with the majority of the people impacted returning to their place of origin. This involuntary and unforeseen movement of people from their place of residence due to extreme weather-related impacts on property and infrastructure poses a challenge for policy-makers and practitioners. Therefore, based on analysis of demographic changes, governance processes have agency in planning for different forms of mobility through creating favourable conditions for voluntary movement as well as prepare for planned relocation alternatives (Adger et al. 2018).

Planned relocation, also termed resettlement or managed retreat, is a structured form of response in the face of sea-level rise or other environmental hazards affecting the lives, livelihood and property (see Chapter 2, Wong et al. 2014; Hino et al. 2017; Mortreux et al. 2018). Relocation is typically initiated, supervised and implemented from national to local level and develops from small communities, but it may also involve large populations (Hino et al. 2017). The strategic relocation of structures and vulnerable communities coupled with the abandonment of land to manage natural hazard risk is a potential adaptation in low-lying zones exposed to the impacts of coastal hazards but has been instigated with varying degrees of success (Hino et al. 2017). In addition to implementation costs, planned relocation initiatives result in a range of cultural, social and psychological losses related to disruptions to sense of place and identity, self-efficacy, and rights to ancestral land and culture (McNamara et al. 2018). Those involved in these processes in coastal areas in Ghana speak for themselves on those social and cultural dimensions of relocation in Box 7.1.

There are significant challenges to governments of enacting fair process and providing frontline services for growing urban populations. Initiatives that seek to enhance the rights of the internally displaced as a direct result of change and empower and protect these populations are being discussed in both policy and academic circles. The legitimacy and efficacy of each individual strategy depends on upholding the principle of fairness of process and ultimately respect for the autonomy of individuals and their decisions of where and how they live (Warner et al. 2013).

Box 7.1 Qualitative interviews with residents in the Volta Delta on diverse impacts on resettlement in coastal villages in Ghana

1. *"Resettlement is an expensive exercise. Actually, there is another sea defence going on at Ada. So government is no more waiting for disaster to happen before they reclaim the land (...) when it's coming (a disaster), we see early, then we claim when it's coming"*. (Quantity surveyor, Public Works Department)
2. *"The other problem some of us have seen in that place is that previously they are closer to the beach for their fishing activities but now it is a distance away from the beach. Now you have to walk a distance unlike previously. That is the only thing and the boulders that they packed over there, if they want to drag the net that is also a small problem over there"*. (Planning officer, Keta Municipal Assembly)
3. *"We are fishermen here so any resettlement consideration must not be far from the sea. We will continue to be fishers because that is the work we know how to do well and depend on to cater for our kids, the place must not be too far from the sea. In addition; since we have children, there must be a school in the place. Every human being will definitely fall sick so there must be a hospital at the place. Once we go fishing, we need to sell the catch so there must be a market as well. All those things must be provided at the new location before any resettlement. But if the four aforementioned facilities are not provided; i.e. the place is not close to the sea, no school for our kids to attend, no market for our wives to trade and hospital for us to get health care when we are ill then it will be difficult for us to agree to a resettlement to any such location"*. (Focus group male participant from a community affected by sea erosion)
4. "We are comfortable here. If the community is secure we are comfortable here. Due to livelihood, employment and so on and so forth we are comfortable here. We can get work to do and the community being secured, we can live here and everything will be okay with us". (Male chief fisherman of a resettled community)

7.5 Conclusion

The Anthropocene is marked by changes that occur at different scales and speeds. Delta communities worldwide, particularly in the developing world, are exposed to major uncertainties to their socio-ecological systems with an element of control for only some risks. The interaction between multiple socio-economic and biophysical changes at different geographical scales and speeds interact to produce uneven outcomes for people's lives and livelihoods in locations exposed to the impacts of this interrelation. As governments design and invest in protective measures and wider coastal development policies, strategies will have major implications for continued settlement of high-risk locations. Current and future migration flows in deltas involve movement of people from unprotected settlements displaced by the slow and continuous loss of productive land due to slow onset processes such as sea-level rise, punctuated by periodic surges of migrants in the aftermath of extreme storm events (McLeman 2018a).

Continued growth of delta urban settlements in itself generates challenges of social cohesion, dependence on hinterlands and population stagnation in non-metropolitan areas. The destinations for migrants are often predictable: existing social networks, historical linkages and economic opportunities tend to be the main pull factors between origin and destination areas. Migration affects social cohesion in destination settlements and communities (Benson and O'Reilly 2009; Skeldon 2014): New migrant populations can feel dislocated from norms and cultures in destination areas, while a strong sense of attachment to communities of origin is shown to strengthen intra-community ties.

Social and environmental change in deltas has the significant potential to disrupt migration flows and individual migration decisions in deltaic areas worldwide (Call et al. 2017). Migratory systems can be altered in multiple ways, ranging from temporary displacement from weather-related disasters to long term decline of regions resulting in planned relocation or managed retreat initiatives. The demographic dimensions of where people live and move in deltas demonstrate how socio-economic transformation and climate change are intertwined with the political economy of development processes. Migration is a successful and desirable option for many people, availing themselves of better lives and opportunities. It is a reality of life in deltas. In effect, it is suggested that the sustainability of delta cities is dependent on the effectiveness and speed of integration of new migrant populations into the social and political life of these places. Investment in the future of cities involves people and place-making as much as infrastructure.

References

Abel, G. J., Barakat, B., Kc, S., & Lutz, W. (2016). Meeting the sustainable development goals leads to lower world population growth. *Proceedings of the National Academy of Sciences, 113*(50), 14294–14299. https://doi.org/10.1073/pnas.1611386113.

Abu, M., Codjoe, S. N. A., & Sward, J. (2014). Climate change and internal migration intentions in the forest-savannah transition zone of Ghana. *Population and Environment, 35*(4), 341–364. https://doi.org/10.1007/s11111-013-0191-y.

Adams, H. (2016). Why populations persist: Mobility, place attachment and climate change. *Population and Environment, 37*(4), 429–448. https://doi.org/10.1007/s11111-015-0246-3.

Adger, W. N., Safra de Campos, R., & Mortreux, C. (2018). Mobility, displacement and migration, and their interactions with vulnerability and adaptation to environmental risks. In R. McLeman & F. Gemenne (Eds.), *Routledge handbook of environmental displacement and migration* (pp. 29–41). London, UK: Routledge.

Amoako-Johnson, F., Hutton, C. W., Hornby, D., Lázár, A. N., & Mukhopadhyay, A. (2016). Is shrimp farming a successful adaptation to salinity intrusion? A geospatial associative analysis of poverty in the populous Ganges–Brahmaputra–Meghna Delta of Bangladesh. *Sustainability Science, 11*(3), 423–439. https://doi.org/10.1007/s11625-016-0356-6.

Anthony, E. J., Almar, R., & Aagaard, T. (2016). Recent shoreline changes in the Volta River Delta, West Africa: The roles of natural processes and human impacts. *African Journal of Aquatic Science, 41*(1), 81–87. https://doi.org/10.2989/16085914.2015.1115751.

Appeaning Addo, K., Nicholls, R. J., Codjoe, S. N. A., & Abu, M. (2018). A biophysical and socioeconomic review of the Volta Delta, Ghana. *Journal of Coastal Research, 34*(5), 1216–1226. http://dx.doi.org/10.2112/JCOASTRES-D-17-00129.1.

BBS. (2011). *Population and housing census 2011*. Bangladesh Bureau of Statistics (BBS), Ministry of Planning, Government of the People's Republic of Bangladesh, Dhaka, Bangladesh. http://www.bbs.dhaka.gov.bd. Last accessed 12 November 2018.

Benson, M., & O'Reilly, K. (2009). Migration and the search for a better way of life: A critical exploration of lifestyle migration. *The Sociological Review, 57*(4), 608–625. https://doi.org/10.1111/j.1467-954X.2009.01864.x.

Bernard, A., Bell, M., & Charles-Edwards, E. (2014). Life-course transitions and the age profile of internal migration. *Population and Development Review, 40*(2), 213–239. https://doi.org/10.1111/j.1728-4457.2014.00671.x.

Bongaarts, J. (2008). Fertility transitions in developing countries: Progress or stagnation? *Studies in Family Planning, 39*(2), 105–110. https://doi.org/10.1111/j.1728-4465.2008.00157.x.

Boyle, P., & Halfacree, K. (Eds.). (2002). *Migration and gender in the developed world*. London, UK: Routledge.

Brown, S., Nicholls, R. J., Lázár, A. N., Hornby, D. D., Hill, C., Hazra, S., et al. (2018). What are the implications of sea-level rise for a 1.5, 2 and 3 °C rise in global mean temperatures in the Ganges-Brahmaputra-Meghna and other vulnerable deltas? *Regional Environmental Change, 18*(6), 1829–1842. http://dx.doi.org/10.1007/s10113-018-1311-0.

Caldwell, J. C. (2006). Demographic theory: A long view. In J. C. Caldwell (Ed.), *Demographic transition theory* (pp. 301–320). Dordrecht, The Netherlands: Springer.

Call, M. A., Gray, C., Yunus, M., & Emch, M. (2017). Disruption, not displacement: Environmental variability and temporary migration in Bangladesh. *Global Environmental Change, 46*, 157–165. https://doi.org/10.1016/j.gloenvcha.2017.08.008.

Census of India. (2011). *Fertility data*. Office of the Registrar General and Census Commissioner, Government of India. http://censusindia.gov.in/. Last accessed 12 November 2018.

Dammers, E., Bregt, A. K., Edelenbos, J., Meyer, H., & Pel, B. (2014). Urbanized deltas as complex adaptive systems: Implications for planning and design. *Built Environment, 40*(2), 156–168.

de Haas, H. (2010). Migration and development: A theoretical perspective. *International Migration Review, 44*(1), 227–264. https://doi.org/10.1111/j.1747-7379.2009.00804.x.

de Jong, G. F., & Fawcett, J. T. (1981). Motivations for migration: An assessment and a value-expectancy research model. In G. F. De Jong & R. W. Gardner (Eds.), *Migration decision making: Multidisciplinary approaches to microlevel studies in developed and developing countries*. New York, NY: Pergamon.

Dyson, T. (2013). On demographic and democratic transitions. *Population and Development Review, 38*(s1), 83–102. https://doi.org/10.1111/j.1728-4457.2013.00553.x.

Ellis, F. (2000). *Rural livelihoods and diversity in developing countries*. Oxford, UK: Oxford University Press.

Ericson, J. P., Vörösmarty, C. J., Dingman, S. L., Ward, L. G., & Meybeck, M. (2006). Effective sea-level rise and deltas: Causes of change and human dimension implications. *Global and Planetary Change, 50*(1–2), 63–82. https://doi.org/10.1016/j.gloplacha.2005.07.004.

GSS. (2013). *National analytical report: 2010 Population and housing census*. Accra, Ghana: Ghana Statistical Service. http://www.statsghana.gov.gh/docfiles/publications/2010_PHC_National_Analytical_Report.pdf. Last accessed 28 August 2018.

Hino, M., Field, C. B., & Mach, K. J. (2017). Managed retreat as a response to natural hazard risk. *Nature Climate Change, 7*(5), 364. http://dx.doi.org/10.1038/nclimate3252.

Hossain, M. S., Uddin, M. J., & Fakhruddin, A. N. M. (2013). Impacts of shrimp farming on the coastal environment of Bangladesh and approach for management. *Reviews in Environmental Science and Bio/Technology, 12*(3), 313–332. https://doi.org/10.1007/s11157-013-9311-5.

Hunter, L. M., Luna, J. K., & Norton, R. M. (2015). Environmental dimensions of migration. *Annual Review of Sociology, 41*(1), 377–397. https://doi.org/10.1146/annurev-soc-073014-112223.

iDMC. (2018). *GRID 2018: Global report on internal displacement*. Geneva, Switzerland: Internal Displacement Monitoring Centre. http://www.internal-displacement.org/sites/default/files/publications/documents/2018-GRID.pdf. Last accessed 12 November 2018.

Levitt, P., & Lamba-Nieves, D. (2011). Social remittances revisited. *Journal of Ethnic and Migration Studies, 37*(1), 1–22. https://doi.org/10.1080/1369183X.2011.521361.

Lutz, W., Butz, W. P., & Samir, K. E. (Eds.). (2017). *World population and human capital in the twenty-first century: An overview*. Oxford, UK: Oxford University Press.

McLeman, R. (2018a). Migration and displacement risks due to mean sea-level rise. *Bulletin of the Atomic Scientists, 74*(3), 148–154. http://dx.doi.org/10.1080/00963402.2018.1461951.

McLeman, R. (2018b). Thresholds in climate migration. *Population and Environment, 39*(4), 319–338. http://dx.doi.org/10.1007/s11111-017-0290-2.

McMichael, A. J. (2014). Population health in the Anthropocene: Gains, losses and emerging trends. *The Anthropocene Review, 1*(1), 44–56. https://doi.org/10.1177/2053019613514035.

McNamara, K. E., Bronen, R., Fernando, N., & Klepp, S. (2018). The complex decision-making of climate-induced relocation: Adaptation and loss and damage. *Climate Policy, 18*(1), 111–117. https://doi.org/10.1080/14693062.2016.1248886.

Mia, M. A., Nasrin, S., Zhang, M., & Rasiah, R. (2015). Chittagong, Bangladesh. *Cities, 48*, 31–41. https://doi.org/10.1016/j.cities.2015.05.011.

Mortreux, C., Safra de Campos, R., Adger, W. N., Ghosh, T., Das, S., Adams, H., et al. (2018). Political economy of planned relocation: A model of action and inaction in government responses. *Global Environmental Change, 50*, 123–132. http://dx.doi.org/10.1016/j.gloenvcha.2018.03.008.

Nicholls, R. J., Brown, S., Goodwin, P., Wahl, T., Lowe, J., Solan, M., et al. (2018). Stabilization of global temperature at 1.5°C and 2.0°C: Implications for coastal areas. *Philosophical Transactions of the Royal Society, 376*(2119). http://dx.doi.org/10.1098/rsta.2016.0448.

Nicholls, R. J., & Cazenave, A. (2010). Sea-level rise and its impact on coastal zones. *Science, 328*(5985), 1517. http://dx.doi.org/10.1126/science.1185782.

Renaud, F. G., Syvitski, J. P. M., Sebesvari, Z., Werners, S. E., Kremer, H., Kuenzer, C., et al. (2013). Tipping from the Holocene to the Anthropocene: How threatened are major world deltas? *Current Opinion in Environmental Sustainability, 5*(6), 644–654. http://dx.doi.org/10.1016/j.cosust.2013.11.007.

Safra de Campos, R., Bell, M., & Charles-Edwards, E. (2016). Collecting and analysing data on climate-related local mobility: The MISTIC Toolkit. *Population, Space and Place, 23*(6), e2037. http://dx.doi.org/10.1002/psp.2037.

Samir, K., & Lutz, W. (2017). The human core of the shared socioeconomic pathways: Population scenarios by age, sex and level of education for all countries to 2100. *Global Environmental Change, 42,* 181–192. https://doi.org/10.1016/j.gloenvcha.2014.06.004.

Seto, K. C. (2011). Exploring the dynamics of migration to mega-delta cities in Asia and Africa: Contemporary drivers and future scenarios. *Global Environmental Change, 21,* S94–S107. https://doi.org/10.1016/j.gloenvcha.2011.08.005.

Seto, K. C., Güneralp, B., & Hutyra, L. R. (2012). Global forecasts of urban expansion to 2030 and direct impacts on biodiversity and carbon pools. *Proceedings of the National Academy of Sciences, 109*(40), 16083. http://dx.doi.org/10.1073/pnas.1211658109.

Skeldon, R. (2014). *Migration and development: A global perspective* (Originally published 1997). London, UK: Routledge.

Stark, O. (1991). Migration in LDCs: Risk, remittances, and the family. *Finance and Development, 28*(4), 39.

Szabo, S., Adger, W. N., & Matthews, Z. (2018). Home is where the money goes: Migration-related urban-rural integration in delta regions. *Migration and Development, 7*(2), 163–179. https://doi.org/10.1080/21632324.2017.1374506.

Szabo, S., Brondizio, E., Renaud, F. G., Hetrick, S., Nicholls, R. J., Matthews, Z., et al. (2016). Population dynamics, delta vulnerability and environmental change: Comparison of the Mekong, Ganges–Brahmaputra and Amazon delta regions. *Sustainability Science, 11*(4), 539–554. http://dx.doi.org/10.1007/s11625-016-0372-6.

Tacoli, C., & Mabala, R. (2010). Exploring mobility and migration in the context of rural—Urban linkages: Why gender and generation matter. *Environment and Urbanization, 22*(2), 389–395. https://doi.org/10.1177/0956247810379935.

te Lintelo, D. J. H., Gupte, J., McGregor, J. A., Lakshman, R., & Jahan, F. (2018). Wellbeing and urban governance: Who fails, survives or thrives in informal settlements in Bangladeshi cities? *Cities, 72,* 391–402. https://doi.org/10.1016/j.cities.2017.10.002.

Warner, K. (2010). Global environmental change and migration: Governance challenges. *Global Environmental Change, 20*(3), 402–413. https://doi.org/10.1016/j.gloenvcha.2009.12.001.

Warner, K., Afifi, T., Kalin, W., Leckie, S., Ferris, B., Martin, S. F., et al. (2013). *Changing climate, moving people: Framing migration, displacement and planned relocation.* UNU-EHS Publication Series, Policy Brief 8. Tokyo, Japan: United Nations University.

Webber, M., & Barnett, J. (2010). *Accommodating migration to promote adaptation to climate change.* Washington, DC: The World Bank.

Whitmee, S., Haines, A., Beyrer, C., Boltz, F., Capon, A. G., de Souza Dias, B. F., et al. (2015). Safeguarding human health in the Anthropocene epoch: Report of The Rockefeller Foundation, Lancet Commission on planetary health. *The Lancet, 386*(10007), 1973–2028. http://dx.doi.org/10.1016/S0140-6736(15)60901-1.

Wong, P. P., Losada, I. J., Gattuso, J.-P., Hinkel, J., Khattabi, A., McInnes, K. L., et al. (2014). Coastal systems and low-lying areas. In C. B. Field, V. R. Barros, D. J. Dokken, K. J. Mach, M. D. Mastrandrea, T. E. Bilir, M. Chatterjee, K. L. Ebi, Y. O. Estrada, R. C. Genova, B. Girma, E. S. Kissel, A. N. Levy, S. MacCracken, P. R. Mastrandrea, & L. L. White (Eds.), *Climate change 2014: Impacts, adaptation, and vulnerability. Part A: Global and sectoral aspects* (pp. 361–409). Contribution of Working Group II to the fifth Assessment Report of the Intergovernmental Panel on Climate Change. Cambridge, UK and New York, NY: Cambridge University Press.

World Bank Data. (2018). *Fertility rate (Bangladesh)*. https://data.worldbank.org/indicator/SP.DYN.TFRT.IN. Last accessed 26 October 2018.

World Bank Group. (2014). *World development indicators 2014*. World Bank Publications. https://data.worldbank.org/products/wdi. Last accessed 9 September 2018.

Zelinsky, W. (1971). The hypothesis of the mobility transition. *Geographical Review, 61,* 219–249. https://doi.org/10.2307/213996.

8

Delta Economics and Sustainability

Iñaki Arto, Ignacio Cazcarro, Anil Markandya, Somnath Hazra, Rabindra N. Bhattacharya and Prince Osei-Wusu Adjei

8.1 Introduction

Deltas are exposed to multiple climatic and environmental factors such as sea-level rise, subsidence, storm surge and temperature and rainfall changes. These environmental risks frequently translate into harm to populations, depending on the sensitivity of the economic activity on which populations depend. Economic characteristics are, of course, heterogeneous across deltas. In absolute terms, the deltas located in East and South-east Asia, Gulf of Mexico, and the deltas of the rivers Rhine, Nile and Parana have a higher GDP per capita compared to the economies in which they are located. The opposite is found in deltas located in less developed regions and in Africa in particular. Developed regions often have significant assets that indicate the financial capacity to cope with environmental change (Tessler et al. 2015). Consequently, regions

I. Arto (✉) · I. Cazcarro · A. Markandya
bc³—Basque Centre for Climate Change, Bilbao, Spain
e-mail: inaki.arto@bc3research.org

R. J. Nicholls et al. (eds.), *Deltas in the Anthropocene*,
https://doi.org/10.1007/978-3-030-23517-8_8

with significant economic activity and infrastructure have been commonly designated as having low vulnerability to environmental risks as investing in infrastructure represents a major element of adaptive capacity (Engle 2011; Tessler et al. 2015).

The economies of deltas are heterogeneous in relation to their structure and evolution. The contribution of agriculture and fisheries to the GDP in deltas located in lower income countries is in general higher than in industrialised countries. Activities such as transportation, tourism, and in some cases oil and gas extraction are present across most deltas to varying extents. While the size and structure of delta economies are, in general, stable, some deltas in emerging economies are growing and diversifying rapidly. This is the case for the Bangladeshi section of the Ganges-Brahmaputra-Meghna (GBM) Delta. Some other delta regions such as the Mekong within Vietnam are, by contrast, shrinking in relation to the economy of the rest of the country.

The evolution of deltas in the Anthropocene era is highly influenced by economic and environmental dynamics at different scales. A significant challenge to deltas is how different potential climate futures affect economic options and pathways and how these in turn affect vulnerability and sustainability, the availability of jobs and livelihoods, and potential population movements in the wider regions. In this chapter

I. Cazcarro
Department of Economic Analysis, Aragonese Agency for Research and Development, Agrifood Institute of Aragon, University of Zaragoza, Saragossa, Spain

S. Hazra
School of Oceanographic Studies, Jadavpur University, Kolkata, India

R. N. Bhattacharya
Department of Economics, Jadavpur University, Kolkata, India

P. O.-W. Adjei
Department of Geography and Rural Development, Kwame Nkrumah University of Science and Technology, Kumasi, Ghana

the current and potential future economies of the GBM, Mahanadi and Volta Deltas are therefore analysed. Possible economic trajectories under climate change through macro-scale economic model and scenario simulations, highlighting both the economic activities and the labour implications of diverse futures, are explored. The chapter also introduces a framework to assess the economic impacts of climate change on individual deltas (as defined in Chapters 2–4) and presents summary findings and conclusions on the future of the deltas in a changing environment.

8.2 The Current Socio-Economic Context of Deltas

Analysing discrete geographical regions within national economies is often restricted by the availability of detailed data on economic activities, not least when some parts of the economy involve informal transactions and labour markets. National statistical institutes of all countries report on a regular basis the main economic aggregates and structure of their economies, as well as some economic indicators at sub-national level. The latter information is often limited and reporting units, provinces or districts, do not necessarily match the geographical area of interest. Furthermore, in some cases such as the GBM Delta, deltas cross national boundaries. While it is desirable to address transnational deltas as a single unit, macroeconomic analysis usually differentiates between the countries' economic systems because they are governed by distinct political, institutional and cultural rules. The circulation of goods, labour and capital are also constrained by geopolitical borders.

The first step to understand the current socio-economic context of deltas consists of compiling relevant information available for the administrative regions within the delta from economic accounts, census and statistics on employment, trade, agriculture and forestry and fishing. This information is used to develop multiregional input-output (MRIO) tables for each delta (see Cazcarro et al. 2018 for methods and data sources). The MRIO table represents the socio-economic structure of the country where it is located, divided into the delta and non-delta parts of the economy. It describes the flows of goods and services between all the individual sectors of the delta and non-delta regions and the use of goods and services by final users, quantified in monetary terms. The distinction between the delta and non-delta is especially relevant for the study of labour and populations movement, since decisions by individuals to relocate across economic sectors and places are primarily driven by differences in the economic conditions of delta and non-delta areas. The MRIO tables are also used to analyse the biophysical and socio-economic conditions and change within deltas. Combining a Computable General Equilibrium (CGE) model with changes in the provision of natural resources, for example, reveals their potential effect on the economies of the deltas and linked regions, and how this in turn affects economic vulnerability and sustainability in these regions.

For the deltas analysed, the MRIO tables reveal the structure of the economies of the delta in relation to the rest of the country. For example, the Bangladesh GBM represents one quarter of the economy of Bangladesh. The Volta represents, by contrast, only 3.5% of the economy of Ghana, while the deltas in India are a small fraction of the national economy. These differences are also reflected in terms of income per capita. In India and Ghana, the GDP per capita (measured in purchasing power parity) is lower in the delta than in the non-delta region suggesting that deltas are economically marginalised and are not growth poles within those countries. In Bangladesh, by contrast, the delta has a GDP per capita higher than the rest of the country. These descriptions are dependent, inevitably, on the presence or absence of large urban centres within the deltas: in the Bangladesh case, the higher GDP per capita is explained by the inclusion of Chattogram (formerly known as Chittagong) and Khulna cities within the region.

Differential income levels in delta and non-deltas areas and the size of the economies of both regions partially explain migration patterns across the deltas, as discussed in Chapter 7. Of the deltas analysed here, the lowest GDP per capita appears in the Volta Delta (USD 1048 compared to USD 1215 per capita nationally); GDP per capita in the Indian deltas is substantially greater being USD 1958 per capita in the Mahanadi Delta and USD 2347 per capita in the Indian section of the GBM (also termed the Indian Bengal Delta [IBD]). In both cases though, the average income per capita in the delta remains lower than for the whole country (USD 3172 per capita). Conversely, in Bangladesh the GDP per capita in the delta (USD 1607) is notably higher than in the whole country (USD 1444 per capita).

The MRIO tables also show the sectoral contribution to GDP. Generally, economies in which a high share of the GDP is linked to sectors whose activity relies on environmental conditions, such as agriculture or fishing, are more likely to suffer the impacts of climate change. Figure 8.1 shows the sectoral structure of the GDP and the employment in the deltas; the tertiary sector is revealed as the main source of income in all deltas. In the three Asian deltas the contribution of services, trade and transportation represents more than half of the total GDP of the economy; in the Volta, while lower, they still contribute 40%. However, the contribution of the primary sector remains significant, ranging between 16 and 29%. The majority of the primary sector contribution corresponds to agriculture, with fisheries making a more

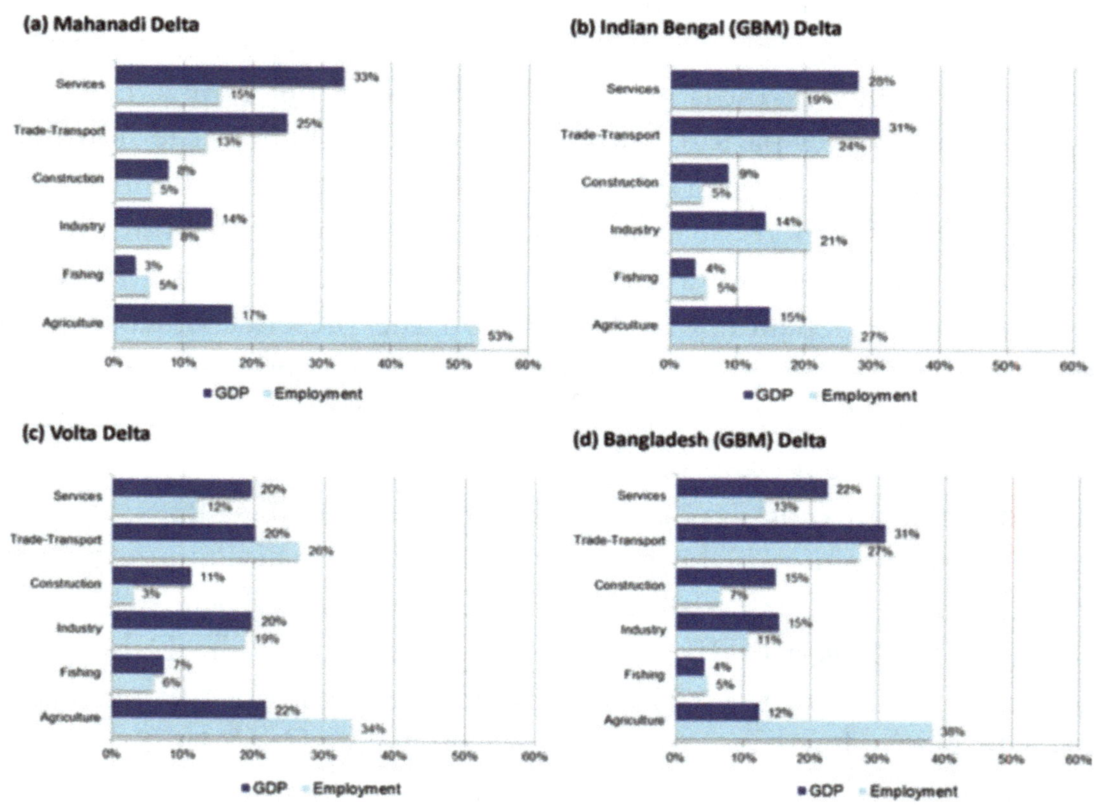

Fig. 8.1 Percentage distribution of GDP and employment by sector and delta in 2011

modest contribution (3–4% of GDP, except in the Volta where it represents 7%). In the Bangladesh GBM the construction sector is relatively high compared to the other delta areas (15%), while industrial activities such as transformation of agricultural goods in the food industry or salt-mining are especially relevant in the Volta (almost 20%). In terms of employment, however, the role of the primary sector in the economy is more relevant than in terms of GDP. The share of the employment engaged in the primary sector ranges from 32% in the IBD to 58% in the Mahanadi. This difference is the consequence of the low productivity of the agricultural sector, in which subsistence production dominates, compared to other economic activities.

In addition to the information shown in Fig. 8.1, the MRIO tables can be used to provide some insights on the importance of each sector in the economy from a systemic view (i.e. including information on the economic activities indirectly linked through supply-chains). In this sense, a way of measuring the relevance of the economic activities exposed to the impacts of climate changes, such as agriculture, is the measurement of the potential impact, in terms of GDP, of their hypothetical disappearance. This impact, which includes both the direct GDP loss in the sector that hypothetically disappears and the cascading or indirect effects in other sectors, is computed using the Hypothetical Extraction Method (Strassert 1968; Meller and Marfán 1981; Cella 1984; Clements 1990). This method hypothetically extracts a sector from an economic system and examines the influence of this extraction on other sectors in the economy. For the Mahanadi Delta, where agriculture is not well-integrated with other sectors, apart from the (hypothetical) direct loss within the sector itself (17% of the GDP), the effect on GDP from other sectors would be relatively low (0.36% of the total Delta GDP). In the IBD direct losses would amount to 15% of the GDP and, as integration is slightly higher, the hypothetical indirect loss in other sectors would be greater (1.23% of the total Delta GDP). In the Bangladesh GBM, direct losses would total 12% of the GDP and indirect would go above 5% of the GDP. The effect is greatest in the Volta Delta where there is a potential direct and indirect reduction in GDP of around 22% and 8% respectively. In the Volta Delta, the relatively higher indirect impact is linked to the relevance of the food processing industry, while in the Bangladesh economy it is due to the food processing industry and textiles and leather transformation. These figures provide a preliminary overview of the vulnerability of the delta economies to impacts such as those originated from climate change which may generate economic losses. However, the comprehensive analysis of the economic impacts of climate change for the future requires further integration between present and future climatic, biophysical and socio-economic drivers under different scenarios.

8.3 Modelling the Economic Impacts of Climate Change in Deltas

Overview of the Integrated Modelling Framework

The analysis of the economic impacts of future climate change in the deltas involves two sources of uncertainty. First, uncertainty arises because of the complexity of functioning of and interactions between complex socio-economic and natural systems. Second, there is intrinsic uncertainty because of the indeterminate nature of future climatic and socio-economic pathways.

Figure 8.2 shows the integrated modelling framework, consisting of a set of scenarios and models operating in different spheres that are used to analyse the impacts of climate change in deltas and to assess different adaptations options, especially migration. The framework follows the typical sequential or cascading structure (Ciscar et al. 2011) linking future climatic projections, changes in environmental conditions, biophysical impacts and economic impacts.

The starting point of the integrated modelling framework is the scenario framework developed by Kebede et al. (2018) (top of Fig. 8.2). This framework uses global narratives for the climatic, socio-economic and adaptation pathways (Representative Concentration Pathways [RCP] [van Vuuren et al. 2011], Shared Socio-economic Pathways [SSP] [O'Neill et al. 2014], and Shared Policy Assumptions [SPA] [Kriegler et al. 2014]). The large-scale global circulation models (GCMs) simulate climate across the World and assess the impacts of

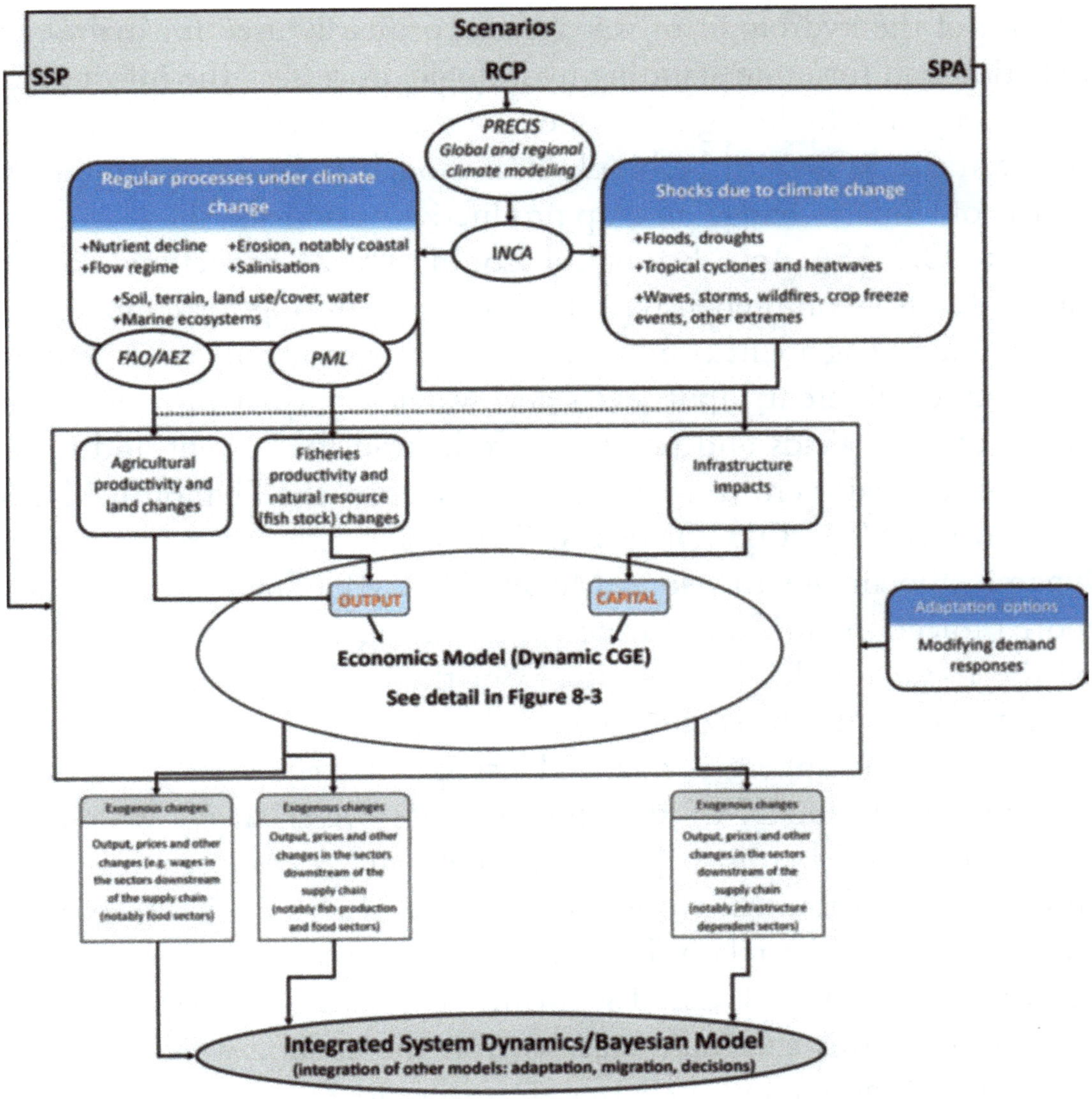

Fig. 8.2 Integrated modelling framework (Adapted from Arto et al. [2019]. Red text indicates link node with Fig. 8.3)

increasing greenhouse gas concentrations on the global climate system. The simulations assume concentrations of greenhouse gases and temperature increases in line with the worst case scenario (i.e. the RCP 8.5). The results of the simulations of various GCMs are downscaled using regional climate models (RCMs).

In the second stage of the modelling chain, the hydrological model takes information from the climate models and provides future pathways for hydrology parameters. Hydrological models such as the INCA model (Whitehead et al. 2015a, b, 2018) are conceptual representations of a part of the hydrologic or water cycle, primarily used for hydrologic prediction and for understanding hydrologic processes. The information reported by this hydrological model is passed, together with information from the RCM, to the biophysical models in order to explore the impacts of climate change on crop productivity and fisheries.

The FAO/IIASA Agro-Ecological Zoning (AEZ) modelling (Fischer et al. 2012) is a comprehensive framework accounting for climate, soil, terrain and management conditions matched with specific crop requirements under different input levels and water supply. It provides information of crop yields and potential production for current and future scenarios for major crops. In the case of fisheries, the Plymouth Marine Laboratory (PML) POLCOMS-ERSEM biogeochemical model and the Dynamic Bioclimate Envelope Model (DBEM) (Cheung et al. 2009) report projections for key marine species using inputs from the RCM and the INCA model. The output from the crop and fisheries models enters as input into the economic model.

The economic sphere is analysed using, for each delta, a dynamic CGE model (Delta-CGE) that interacts at several stages with other models. The model consists of two components: a comprehensive economic dataset of the case study areas assembled in a Social Accounting Matrix (SAM), and a relatively flexible model (systems of equations) in the form of a dynamic CGE. The former represents the flows of all economic transactions that take place within the economy (an extension of the MRIO tables) and the latter aims to represent the flows of goods and services in the economies of the deltas and their relation with the rest of the country at a given point in time. In this regard, the Delta-CGE model uses actual economic data to estimate how the economy might react to changes in external factors.

The starting point of the economic modelling are three scenario storylines up to 2050 inspired by the SSP2, SSP3 and SSP5 narratives (Moss et al. 2010; van Vuuren et al. 2011; O'Neill et al. 2012; Kriegler et al. 2014; Riahi et al. 2017). At the national scale, the socio-economic scenarios for the three countries (Ghana, India and Bangladesh) are based on the publicly available SSPs (IIASA, 2018). These data provide historic trends and future projections of the changes in urban and rural populations, and GDP for each country under the different SSP scenarios. These data are used as one of the boundary conditions to develop the scenario at the delta-level in collaboration with regional and national experts and stakeholders.

The Delta-CGE model (Arto et al. 2019) acts as an interface between the climate and biophysical models and the integrated model of migration, in the sense that it translates the biophysical impacts of climate change, such as reduction of crop productivity, into changes in some key socio-economic drivers of migration, such as wages. It is important to highlight that the Delta-CGE model does not seek to directly translate changes in climatic conditions into migration flows. Rather, it aims at taking advantage of the superiority of the biophysical models to capture the impacts of climate change in some critical variables affecting specific economic processes and translates them into economic impacts.

The economic model also interacts with the four different adaptation trajectories elaborated through combining expert-based and participatory methods. The four adaptation trajectories are termed Minimum Intervention, Economic Capacity Expansion, System Efficiency Enhancement and System Restructuring (see Suckall et al. 2018 for further details). The combination of the three different socio-economic scenarios, the four adaptation trajectories and the impacts of climate change from the biophysical models results on a set of possible socio-economic trajectories under climate change.

This information is further passed to the Integrated System Dynamics models and Bayesian Network model (see Lázár et al. 2017) where, in combination with the outputs of other models, it is used to assess the impact of climate change on human well-being and to evaluate different coping strategies. The integration of the results of the economic model into the migration models is done through a statistical emulator, which approximates the results through statistical relationships based on a Monte Carlo analysis of the economics results, without the need of integrating all the sets of equations of the Delta-CGE. At the same time, partial assessments of these integrated models provide the Delta-CGE with an ex-ante exogenous default set of migration figures (without yet accounting for migration changes estimated in the economics model), represented as dotted lines in Fig. 8.2.

The Delta-CGE Model

The Delta-CGE shows a set of features that make it appropriate to assess migration as an adaptation in deltaic environments under a changing climate and to inform sustainable gender-sensitive adaptation (see Fig. 8.3).

The production side of the economy is defined by production functions which specify, for each sector, the inputs required to produce one unit of output. These inputs can be intermediate inputs (goods and services from other sectors, which may come from the delta, the rest of the country or the rest of the World), as well as production factors: capital, land and labour. The level of capital depends on the capital of the previous year and current investment, which is influenced by interest rates and depreciation rates. Capital stock and investment are also affected by climate change through losses due to extreme events (e.g. damages in infrastructures). Land is used in the agriculture sectors to produce different crops and can be directly affected by climate change through losses in land availability. In addition, the output of agricultural products can also be affected by climate change through changes in crop yield.

Labour is linked to population, which determines the labour force, and to households, which receive the income from labour compensation

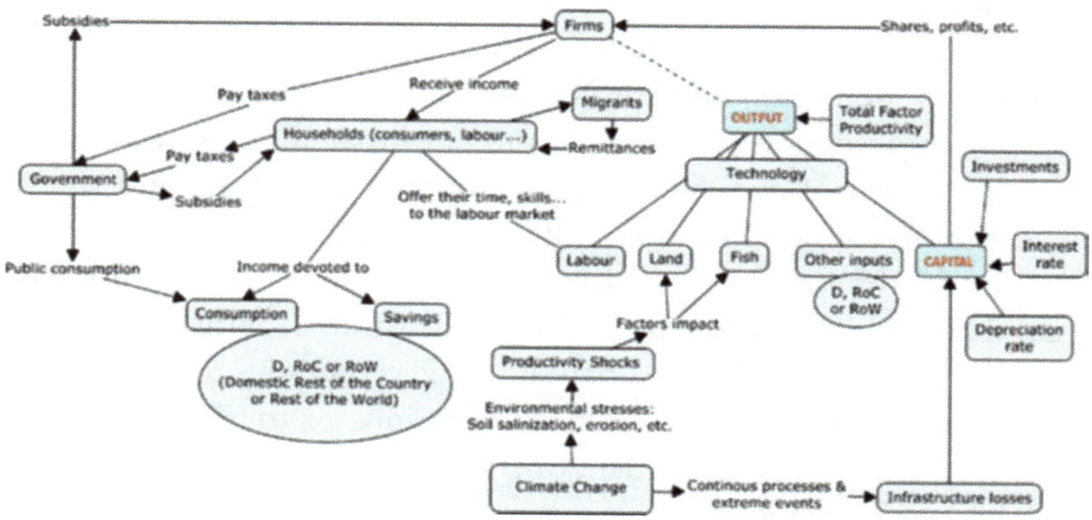

Fig. 8.3 General structure of the Delta-CGE model (Red text indicates link node with Fig. 8.2)

(i.e. wages). Households consume goods and services coming from the delta, the rest of the country or the rest of the World. Income is distributed between consumption and savings/investments. Households also interact with government by paying taxes and receiving transfers. A similar interaction applies to firms and government. It is also via firms that the returns of capital are further invested within companies or distributed to households. The income of households in the delta also includes remittances from migrants in other regions.

The model also has a migration module in which migration is driven by economic factors using a Harris-Todaro model type of decision rule (Harris and Todaro 1970; Todaro 1986; Gupta 1993; Espíndola et al. 2006). According to this rule, migration is triggered in response to differences in expected earnings between regions. The specification may look oversimplified, but other factors affecting migration are modelled in the Integrated System Dynamics models and Bayesian Network model (Lázár et al. 2017) and linked to the Delta-CGE model. The dominant thinking in the economics literature of labour migration highlights how better economic opportunities typically drive migration; this has been found in the deltas analysed (Safra de Campos et al. 2016) (see Chapter 7). Arto et al. (2019) provide a detailed description of the migration module of the Delta-CGE model.

8.4 The Socio-Economic Future of Deltas in a Changing Environment

Baseline Socio-Economic Scenarios

Socio-economic scenario modelling, together with climate and environmental analyses, is one of the key elements in the study of climate change and its possible implications in the mid-term. Several approaches have been taken in the literature, such as Participatory Scenario Development (e.g. Bizikova et al. 2014), also sometimes linking stakeholder survey, scenario analysis, and simulation modelling to explore long-term trajectories (e.g. Keeler et al. 2015). The scenario framework (see Kebede et al. 2018), integrates knowledge from scenario design, modelling and surveys, being stakeholder participation a central element of the framework. This was done with the triple purpose of engaging stakeholders in the project, understanding the capacity of the governance system to support migration in the context of other adaptation options, and leaving a policy and practical legacy from the research.

The socio-economic projections also include a modest exercise of expert-based questionnaire about the future of the deltas regarding rural/urban population, GDP growth and composition, inequality, or education. These expert insights are treated with caution, and put in context in relation to other knowledge, literature and complementary analyses as benchmarks for comparison. In particular, key reference data at the country level are the GDP and population levels projected by the SSP Public Database Version 1.1 (see Riahi et al. 2017). Looking at the future GDP growth (in purchasing power parities, PPP) (see Fig. 8.4), for Ghana the experts' views lie in between the SSP2 and SSP5 projections. In India, experts envisage a lower GDP growth path for deltas than those of SSP3. However, for Bangladesh (both for the delta and non-delta) the experts' visions on GDP are quite above the scenario of the highest growth (i.e. SSP5). This becomes similarly clear when looking at the projected GDP per capita, which would reach close to USD 8000 per year in Bangladesh under the high growth scenario. In the case of Ghana, also the high growth scenario implies even clearer increases in GDP per capita, reaching over USD 8000 per year, due to the clear cut slower population growth under this scenario. In the case of India, the point of departure is much higher; the experts' views from the questionnaires are relatively less pro-growth than in other deltas, being around the projections from SSP3. Still, in those lower-case options the projections are of a GDP per capita around USD 11,000 going up to around USD 30,000 per capita in the SSP5.

As with all projections, these figures should be taken with care. The responses to the questionnaires are sometimes too "pro-GDP growth optimistic", as witnessed in past literature, and are often based on early trends which then were proven to deviate downward, due to the fact that, for example, continuous GDP growth has rarely been sustained for

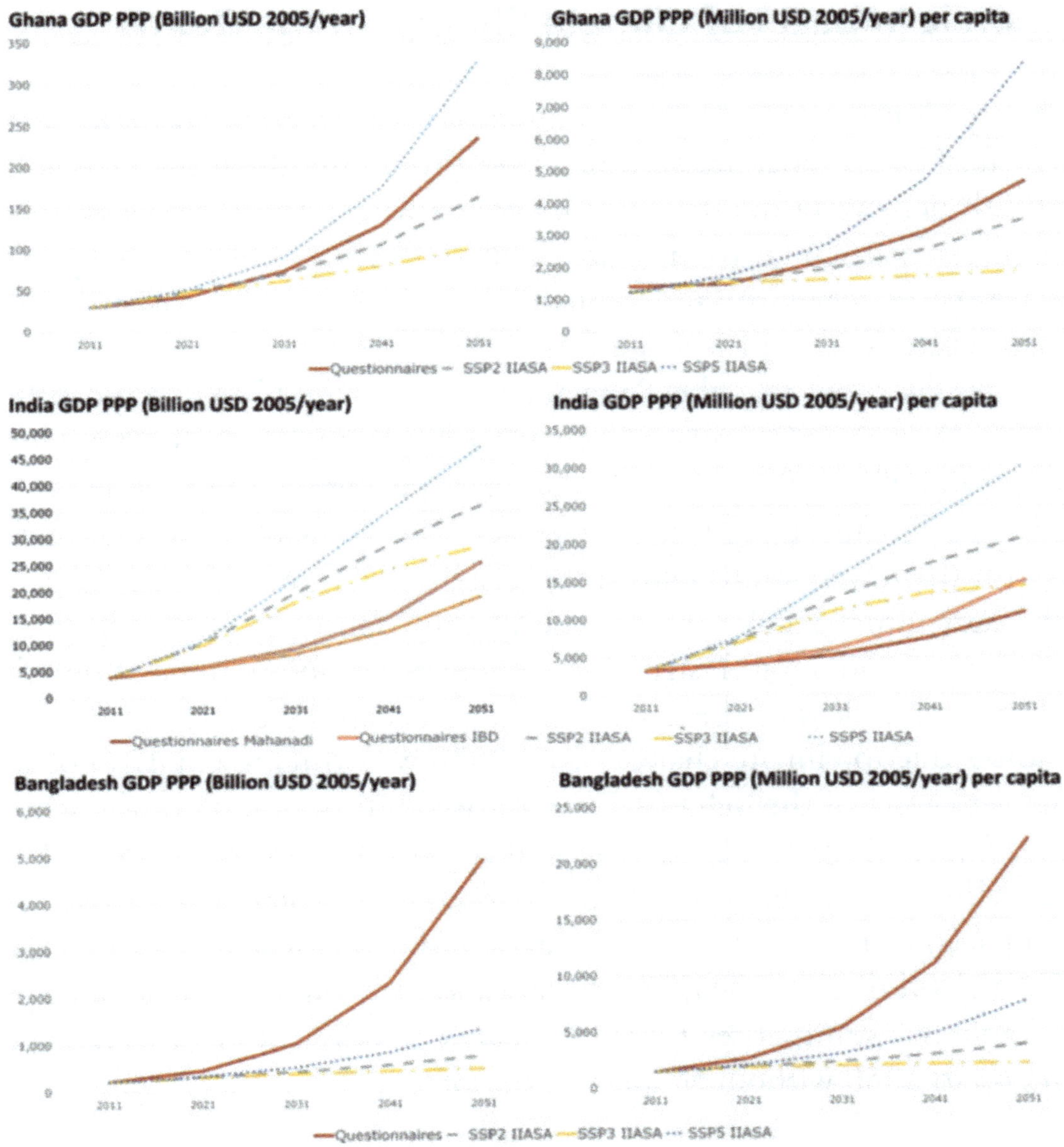

Fig. 8.4 Country GDP (PPP) and GDP (PPP) per capita projected by SSPs (IIASA) and delta questionnaires (please note differences in vertical scales)

long periods for any country. On the other hand, it is true that SSP projections for Bangladesh on GDP, and more importantly on GDP per capita, seem to be low compared to current trends, so these types of paths are also worth considering, especially if they are likely to be associated to other processes of urbanisation, migration, environmental degradation, etc.

Economic Impacts of Climate Change in the Deltas

In this section the outcomes of simulations for different scenarios with the Delta-CGE model are summarised. The results show the change in the GDP per capita due to climate change with respect to the baseline scenario. In particular, the impacts analysed include shocks on agriculture (losses in terms of land availability and crop yield), fisheries and infrastructure, with and without adaptation options.

The results show that the Bangladeshi side of the GBM would suffer the highest economic impact from climate change, with a cumulative loss up to 2050 of 19.5% in terms of GDP per capita, mostly due to the impacts in agriculture and infrastructures. The Volta Delta would be the one with the lowest GDP per capita losses (−9%), mainly from impacts in the agriculture sector. In the case of the Mahanadi Delta, the main shocks are found in infrastructure, representing about three times the GDP per capita loss of the agricultural and fisheries sectors; the cumulative losses due to climate change up to 2050 would represent 12% of the GDP per capita of the delta (0.25% of the GDP per capita of the whole India). Finally, in the IBD, damages in infrastructure up to 2050 would generate losses equivalent to 7% of the GDP per capita, losses driven by the impacts of climate change in the agricultural sector would affect about 8% of the GDP per capita, and fisheries losses would clearly stay below 1%. Table 8.1 summarises these cumulative losses in terms of GDP per capita up to 2050 without adaptation options.

Table 8.1 Cumulative GDP per capita percentage losses due to climate change by type of impact in selected deltas up to 2050, average and range assuming no adaptation

	GBM		Volta	Mahanadi
	Bangladesh	IBD		
Agriculture[a]	12 (8–14)	8 (6–9)	6.5 (3–9)	3 (1–6)
Fisheries[b]	0.36	0.33	0.85	0.09
Infrastructure	7.5 (6–9)	7 (4.5–9)	1.5 (1–2)	9 (7–11)
Total	19.5 (14–24)	15.5 (10–19)	9 (4–13)	12 (8–16)

[a]Conditioned on whether CO_2 fertilisation and good management practices take place or not

[b]Average of three climatic models implemented with the fisheries modelling POLCOMS-ERSEM

Adaptation options, in particular embankments and protection and restoration of mangroves would contribute to buffer these effects. In the case of the Mahanadi Delta, losses in terms of GDP per capita could be reduced by 2%. For the IBD, from the 15% reduction in terms of GDP per capita due to climate change, about 2% could be buffered with adaptation interventions (Fig. 8.5).

In the case of the Volta Delta, the effects from the expected climate impacts on fisheries are the most important across all studied deltas (9–17% of decrease in potential catch up to 2100, and 4–8% in 2050), implying around 1% GDP per capita loss up to mid-century. This could be buffered by adaptation activities such as housing for fishing communities, establishment of fish seed hatcheries and further development of retail fish markets and allied infrastructure. The Bangladesh GBM is a delta without a development gap with respect to the rest of the country and with lower specialisation in agriculture. However, the impacts of climate change in the

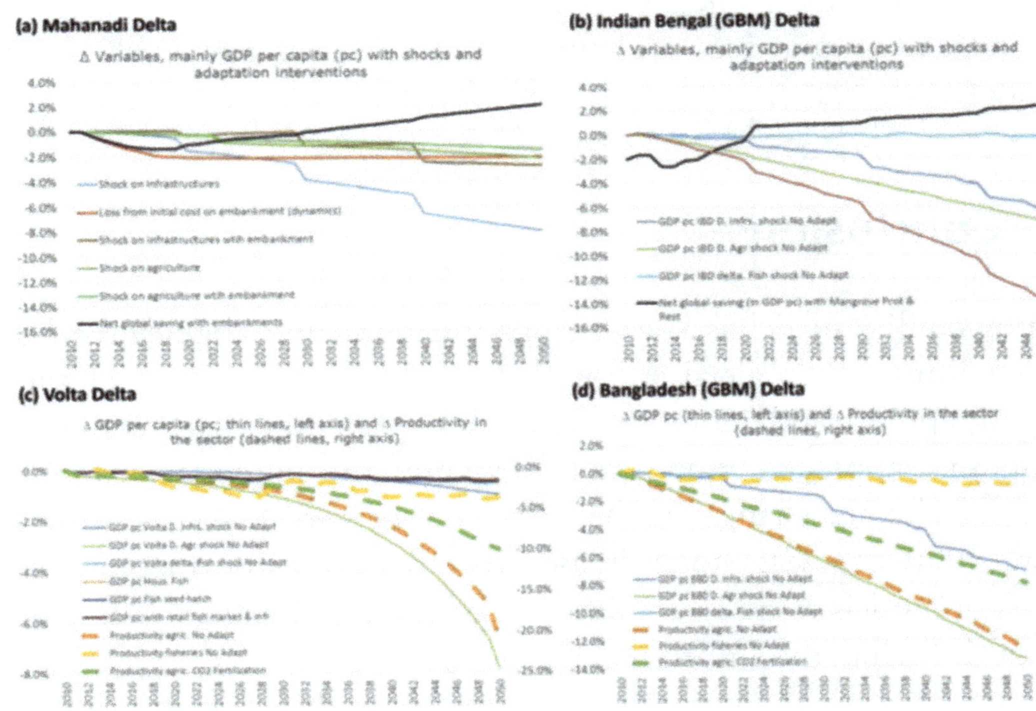

Fig. 8.5 Economic impacts of climate change for the four delta economies. Percentage change with respect to the baseline scenario

agricultural sector are still the highest of the areas analysed, reaching a reduction of about 12% GDP per capita in 2050. This result arises from the expected reduction in crop yield that may reach 30% by 2050, and even with CO_2 fertilisation and good management practices still could reach 20%. This decline in yield also translates into GDP per capita reductions for the rest of the country (non-delta region) of about 2% in 2050.

As previously indicated, the socio-economic analysis of fisheries impacts builds on fisheries modelling for both the Gulf of Guinea and Bay of Bengal (Fernandes et al. 2016; Lauria et al. 2018). The expected impacts of climate change on fisheries up to the year 2050 are entered as input in the Delta-CGE model as fisheries losses (based on likeliness of fisheries changes, which may involve growth of stock of some species, and higher losses in others) for each deltaic region. In 2050 under current management practices losses are estimated to be about 8% for Ghana and 4% for the Bay of Bengal. This implies losses in the whole GDP per capita of the deltas of about 0.1% for the Mahanadi Delta, 0.35% for the whole GBM Delta and 0.85% for the Volta Delta. This potential fish decline could be reduce to a third mitigated if sustainable management practices are undertaken (see Barange et al. 2014; Fernandes et al. 2016).

8.5 Conclusions

The vulnerability of deltas is a complex phenomenon characterised by many environmental, social and economic drivers interacting at multiple geographical and temporal scales. Using an integrated biophysical-economic approach, this chapter explores the potential economic impacts of climate change in the deltas of the GBM, Mahanadi and Volta to 2050. A CGE model is used together with information from biophysical models to assess the impacts of climate change on the economies of the deltas and linked regions.

The results from the model simulations show the economic importance of these deltas. Assuming no adaptation, losses in terms of GDP per capita range from an average of 9% in the Volta Delta to 19.5%

in the Bangladeshi section of the GBM. These impacts mainly constitute damages to infrastructure and losses in crop production and, to a lesser extent, due to losses in fisheries. For all these deltas, impacts of agriculture and fisheries represent key livelihoods for the poorest and more vulnerable, and hence the consequences of these purely macroeconomic impacts are expected to be large in terms of livelihoods and food security.

The simulations also provide key information for the development and implementation of successful adaptation options in deltas (Chapters 9 and 10). For example, impacts on agriculture in all the deltas analysed represent around 8% decline in production, though such impacts could be potentially halved through effective management practices. In addition, the results also show the potential effect of interventions aimed at reducing disaster risks such as building multipurpose cyclone shelters or constructing embankments in terms of avoiding economic and social costs of extreme events. This macro-economic analysis and diverse studies of economic behaviour and investments suggest that sustaining people and livelihoods in place in delta regions involves significant challenges, but also offers multiple benefits to the regions and countries within which they reside.

References

Arto, I., García-Muros, X., Cazcarro, I., González, M., Markandya, A., & Hazra, S. (2019). The socioeconomic future of deltas in a changing environment. *Science of the Total Environment, 648,* 1284–1296. https://doi.org/10.1016/j.scitotenv.2018.08.139.

Barange, M., Merino, G., Blanchard, J. L., Scholtens, J., Harle, J., Allison, E. H., et al. (2014). Impacts of climate change on marine ecosystem production in societies dependent on fisheries. *Nature Climate Change, 4*(3), 211. http://dx.doi.org/10.1038/nclimate2119.

Bizikova, L., Pinter, L., & Tubiello, F. N. (2014). *Recent progress in applying participatory scenario development in climate change adaptation in developing countries Part II* (Working Paper). Winnipeg, MB, Canada: International Institute for Sustainable Development (IISD).

Cazcarro, I., Arto, I., Hazra, S., Bhattacharya, R. N., Adjei, P. O.-W., Ofori-Danson, P. K., et al. (2018). Biophysical and socioeconomic state and links of deltaic areas vulnerable to climate change: Volta (Ghana), Mahanadi (India) and Ganges-Brahmaputra-Meghna (India and Bangladesh). *Sustainability, 10*(3), 893. http://dx.doi.org/10.3390/su10030893.

Cella, G. (1984). The input-output measurememt of inter-industry linkages. *Oxford Bulletin of Economics and Statistics, 46*(1), 73–84. https://doi.org/10.1111/j.1468-0084.1984.mp46001005.x.

Cheung, W. W. L., Lam, V. W. Y., Sarmiento, J. L., Kearney, K., Watson, R., & Pauly, D. (2009). Projecting global marine biodiversity impacts under climate change scenarios. *Fish and Fisheries, 10*(3), 235–251. https://doi.org/10.1111/j.1467-2979.2008.00315.x.

Ciscar, J.-C., Iglesias, A., Feyen, L., Szabó, L., Van Regemorter, D., Amelung, B., et al. (2011). Physical and economic consequences of climate change in Europe. *Proceedings of the National Academy of Sciences, 108*(7), 2678–2683.

Clements, B. J. (1990). On the decomposition and normalization of inter-industry linkages. *Economics Letters, 33*(4), 337–340. https://doi.org/10.1016/0165-1765(90)90084-E.

Engle, N. L. (2011). Adaptive capacity and its assessment. *Global Environmental Change, 21*(2), 647–656. https://doi.org/10.1016/j.gloenvcha.2011.01.019.

Espíndola, A. L., Silveira, J. J., & Penna, T. J. P. (2006). A Harris-Todaro agent-based model to rural-urban migration. *Brazilian Journal of Physics, 36,* 603–609.

Fernandes, J. A., Kay, S., Hossain, M. A. R., Ahmed, M., Cheung, W. W. L., Lazar, A. N., et al. (2016). Projecting marine fish production and catch potential in Bangladesh in the 21st century under long-term environmental change and management scenarios. *ICES Journal of Marine Science, 73*(5), 1357–1369. http://dx.doi.org/10.1093/icesjms/fsv217.

Fischer, G., Nachtergaele, F. O., Prieler, S., Teixeira, E., Toth, G., van Velthuizen, H., et al. (2012). *Global Agro-Ecological Zones (GAEZ v3.0)—Model documentation*. Laxenburg, Austria: IIASA; Rome, Italy: FAO.

Gupta, M. R. (1993). Rural-urban migation, informal sector and development policies A theoretical analysis. *Journal of Development Economics, 41*(1), 137–151. https://doi.org/10.1016/0304-3878(93)90040-T.

Harris, J. R., & Todaro, M. P. (1970). Migration, unemployment and development: A two-sector analysis. *The American Economic Review, 60*(1), 126–142.

Kebede, A. S., Nicholls, R. J., Allan, A., Arto, I., Cazcarro, I., Fernandes, J. A., et al. (2018). Applying the global RCP–SSP–SPA scenario framework at sub-national scale: A multi-scale and participatory scenario approach. *Science of the Total Environment, 635*, 659–672. http://dx.doi.org/10.1016/j.scitotenv.2018.03.368.

Keeler, L. W., Wiek, A., White, D. D., & Sampson, D. A. (2015). Linking stakeholder survey, scenario analysis, and simulation modeling to explore the long-term impacts of regional water governance regimes. *Environmental Science and Policy, 48*, 237–249. https://doi.org/10.1016/j.envsci.2015.01.006.

Kriegler, E., Edmonds, J., Hallegatte, S., Ebi, K. L., Kram, T., Riahi, K., et al. (2014). A new scenario framework for climate change research: The concept of shared climate policy assumptions. *Climatic Change, 122*(3), 401–414. http://dx.doi.org/10.1007/s10584-013-0971-5.

Lauria, V., Das, I., Hazra, S., Cazcarro, I., Arto, I., Kay, S., et al. (2018). Importance of fisheries for food security across three climate change vulnerable deltas. *Science of the Total Environment, 640–641*, 1566–1577. http://dx.doi.org/10.1016/j.scitotenv.2018.06.011.

Lázár, A. N., Adams, H., & Safra de Campos, R. S. (2017, September 25–29). *Migration as an adaptation to climate change in deltaic regions: A modelling study*. Social Simulation Conference 2017. Dublin, Ireland: RTI International.

Meller, P., & Marfán, M. (1981). Small and large industry: Employment generation, linkages, and key sectors. *Economic Development and Cultural Change, 29*(2), 263–274. https://doi.org/10.1086/451246.

Moss, R. H., Edmonds, J. A., Hibbard, K. A., Manning, M. R., Rose, S. K., van Vuuren, D. P., et al. (2010). The next generation of scenarios for climate change research and assessment. *Nature, 463*, 747. http://dx.doi.org/10.1038/nature08823.

O'Neill, B. C., Carter, T. R., Ebi, K. L., Edmonds, J., Hallegatte, S., Kemp-Benedict, E., et al. (2012, November 2–4). *Workshop on the nature and use of new socioeconomic pathways for climate change research* (Meeting Report) (pp. 1–37). Boulder, CO. https://www2.cgd.ucar.edu/sites/default/files/iconics/Boulder-Workshop-Report.pdf. Last accessed 4 September 2018.

O'Neill, B. C., Kriegler, E., Riahi, K., Ebi, K. L., Hallegatte, S., Carter, T. R., et al. (2014). A new scenario framework for climate change research: The concept of shared socioeconomic pathways. *Climatic Change, 122*(3), 387–400. http://dx.doi.org/10.1007/s10584-013-0905-2.

Riahi, K., van Vuuren, D. P., Kriegler, E., Edmonds, J., O'Neill, B. C., Fujimori, S., et al. (2017). The shared socioeconomic pathways and their energy, land use, and greenhouse gas emissions implications: An overview. *Global Environmental Change, 42*(Suppl. C), 153–168. http://dx.doi.org/10.1016/j.gloenvcha.2016.05.009.

Safra de Campos, R., Bell, M., & Charles-Edwards, E. (2016). Collecting and analysing data on climate-related local mobility: The MISTIC toolkit. *Population, Space and Place, 23*(6), e2037. http://dx.doi.org/10.1002/psp.2037.

Strassert, G. (1968). Zur Bestimmung strategischer Sektoren mit Hilfe von Input-Output-Modellen. *Jahrbucher Fur Nationalokonomie Und Statistik, 182*(3), 211–215.

Suckall, N., Tompkins, E. L., Nicholls, R. J., Kebede, A. S., Lázár, A. N., Hutton, C., et al. (2018). A framework for identifying and selecting long term adaptation policy directions for deltas. *Science of the Total Environment, 633*, 946–957. http://dx.doi.org/10.1016/j.scitotenv.2018.03.234.

Tessler, Z., Vörösmarty, C. J., Grossberg, M., Gladkova, I., Aizenman, H., Syvitski, J., et al. (2015). Profiling risk and sustainability in coastal deltas of the world. *Science, 349*(6248), 638–643. http://dx.doi.org/10.1126/science.aab3574.

Todaro, M. P. (1986). Internal migration and urban employment: Comment. *The American Economic Review, 76*(3), 566–569.

van Vuuren, D. P., Edmonds, J., Kainuma, M., Riahi, K., Thomson, A., Hibbard, K., et al. (2011). The representative concentration pathways: An overview. *Climatic Change, 109*(1), 5. http://dx.doi.org/10.1007/s10584-011-0148-z.

Whitehead, P. G., Barbour, E., Futter, M. N., Sarkar, S., Rodda, H., Caesar, J., et al. (2015a). Impacts of climate change and socio-economic scenarios on flow and water quality of the Ganges, Brahmaputra and Meghna (GBM) river systems: Low flow and flood statistics. *Environmental Science: Processes Impacts, 17*(6), 1057–1069. http://dx.doi.org/10.1039/C4EM00619D.

Whitehead, P. G., Sarkar, S., Jin, L., Futter, M. N., Caesar, J., Barbour, E., et al. (2015b). Dynamic modeling of the Ganga river system: Impacts of future climate and socio-economic change on flows and nitrogen fluxes in India and Bangladesh. *Environmental Science: Processes and Impacts, 17*(6), 1082–1097. http://dx.doi.org/10.1039/C4EM00616J.

Whitehead, P. G., Jin, L., Macadam, I., Janes, T., Sarkar, S., Rodda, H. J. E., et al. (2018). Modelling transboundary impacts of RCP 8.5 climate change and socio-economic change on flow and water quality of the Ganga, Brahmaputra, Meghna, Hooghly and Mahanadi river systems in India and Bangladesh. *Science of the Total Environment, 636*, 1362–1372. http://dx.doi.org/10.1016/j.scitotenv.2018.04.362.

9

Adapting to Change: People and Policies

Emma L. Tompkins, Katharine Vincent, Natalie Suckall, Rezaur Rahman, Tuhin Ghosh, Adelina Mensah, Kirk Anderson, Alexander Chapman, Giorgia Prati, Craig W. Hutton, Sophie Day and Victoria Price

9.1 Introduction

Deltas are shifting, subsiding, morphing environments endlessly adapting to changes in sediment flows, water levels, storms, floods and sea-level rise, both naturally and, increasingly, through human effort (e.g. Chapters 2–4). Human activity and settlements within deltas and their watersheds

E. L. Tompkins (✉) · N. Suckall · G. Prati · S. Day · V. Price
Geography and Environmental Science,
University of Southampton, Southampton, UK
e-mail: E.L.Tompkins@soton.ac.uk

K. Vincent
Kulima Integrated Development Solutions, Pietermaritzburg, South Africa

R. Rahman
Institute of Water and Flood Management, Bangladesh University of Engineering and Technology, Dhaka, Bangladesh

T. Ghosh
School of Oceanographic Studies, Jadavpur University, Kolkata, India

R. J. Nicholls et al. (eds.), *Deltas in the Anthropocene*,
https://doi.org/10.1007/978-3-030-23517-8_9

contribute to the vulnerable environment within which deltas produce food, support commerce and residents manage their lives and livelihoods. Delivering secure places while improving ability to adapt and fostering resilience is a huge challenge in rapidly changing delta landscapes in the Anthropocene.

Deltas have been changed by human activity since early human settlement. For example, human modifications in the Ganges-Brahmaputra-Meghna dating are documented for approximately 3000 years, and more recently with the founding of Dhaka in 1604 (Fergusson 1863). The Anthropocene is characterised by a great acceleration in trends of land use and other change. Dhaka for example, has increased in population from around 220,000 in 1941 to 15 million in 2011 (RAJUK 2015). During this period the city has expanded with land reclaimed and more low lying flood-prone areas have been settled.

The abundance of fertile land means that deltas are vital resources in food production. However the context in which this takes place in the Anthropocene is changing. As a result of population increase and demand for land, land tends to be used more intensively. Large engineered interventions are common which can include upstream dams outside the delta to generate hydropower on the Nile and Volta Rivers, and canalisation for irrigation and transport as seen within the Mississippi and other deltas. The Mekong Delta, for example, is central to the rice bowl of Southeast

A. Mensah
Institute for Environment and Sanitation Studies, University of Ghana, Legon-Accra, Ghana

K. Anderson
Regional Institute for Population Studies, University of Ghana, Legon-Accra, Ghana

A. Chapman
School of Engineering, University of Southampton, Southampton, UK

C. W. Hutton
GeoData Institute, Geography and Environmental Science, University of Southampton, Southampton, UK

Asia and generates around 50% of Vietnam's total rice output and about 90% of its rice export (Ling et al. 2015). Other deltas, such as the Mahanadi in India, have larger proportions involved in small scale and subsistence farming, representing a significant labour and livelihood for extensive populations (Duncan et al. 2017).

Here, the building blocks of adaptation to environmental change in deltas and prospects for the future are examined. The focus is on decisions made by people, not just as individual agents but also in the social context of households. Such decisions are constrained and shaped by collective and policy-driven adaptation. This chapter considers the lived reality and social distribution of vulnerability and reviews evidence on where adaptation is occurring, who is undertaking it, what forms it takes, and what types of adaptation are perceived to be successful. An adaptation typology to organise forms of adaptation is presented which considers the relationship between policy driven adaptation and what households are doing within this adaptation policy context. Adaptation policy has, on occasion, unforeseen negative consequences of adaptation policy and the chapter reflects on the future of adaptation, specifically the relationship between latent and active capacity to adapt, vulnerability and the existence of incentives to adapt.

9.2 Vulnerability Affects People's Ability to Adapt in Deltas

People in deltas are, in many places, highly vulnerable to environmental shocks and stresses. Many elements of this vulnerability are driven by the natural geography of deltas, e.g. river flow and sedimentation, but are amplified by more recent human interference with the delta systems. This includes inappropriate or poorly maintained engineering interventions, such as dams, navigation, flood control works, but also from demographic pressures and changes in land use. The combination of all of these pressures leads to floods, subsidence, storm surges and a highly variable living environment. Deltas also face upstream and externally driven stresses, such as sediment starvation from dams (Chapter 5), price fluctuations in key crops from global economic issues, and the hazards associated with climate change (Nicholls et al. 2007).

In terms of current vulnerability, the role of sea-level rise remains uncertain. Some argue that present day societal vulnerability is more dependent on risks from river discharge and storm surges, rather than longer term trend changes in sea level (Vermaat and Eleveld 2013). Others argue that sea-level rise and climate change are dominant factors shaping deltaic environments in the future (Szabo et al. 2016). There is no debate that climate change will have an impact on the vulnerability of deltas. The questions to be asked are: is climate change already affecting deltas, if not, when will it start to have an impact, and what can be done about this now?

Levels of economic development play a key role in shaping present day vulnerability (Tessler et al. 2015). Chapter 5, for example, shows how shocks to the regional economies of deltas result in reductions in labour demand, aggregate income levels, and ultimately undermine the resilience of these areas. Deltas in wealthy countries, such as those of the Mississippi and the Rhine, appear to be better placed to cope with current stresses than those in poorer countries, due to levels of investment in protective infrastructure. The distribution of resources, and levels of inequality and poverty, especially in the developing world, make delta populations vulnerable and fragile in the context of environmental shocks. In the Yellow River Delta, China, for example, low levels of education, below minimum wage and general lack awareness of global climatic issues of its many deltaic residents, are considered important factors that contribute to increasing their vulnerability (Wolters et al. 2016). As outlined in Chapter 7, delta areas are characterised by trends towards ageing populations and significant shifts in populations from rural to urban areas (Szabo et al. 2016). At present some rural areas continue to have labour surpluses, but are increasingly facing the implications of an ageing population with high dependency ratios with outfluxes of working age adults to cities.

Even within deltas, experiences vary with the social factors that shape vulnerability and adaptive capacity in deltas (see Chapter 6). Limited access to resources, low decision-making power and social roles constrain women's capacity to prepare, respond and adapt to climate shocks and stresses (Pearse 2016). The adverse impacts of coastal erosion on land and water have gendered effects linked to social responsibilities and roles. Water salinisation and land loss can force women to walk longer distances to collect water and graze livestock adding further physical and time burden that ultimately affects their adaptive capacity. However, it is worth noting that vulnerabilities are not homogenous among women, but are determined by an interplay of social, economic and cultural factors including age, class, caste and ethnicity (Kaijser and Kronsell 2014). A case study in Odisha, shows that women from upper castes are less vulnerable to cyclones than women from low-castes due to better access to social networks, assets and resources (Ray-Bennett 2009). Age can also be a mitigating factor of vulnerability linked to voice and decision-making in intra-households power dynamics.

Perceptions of risk affect vulnerability and are subjective, reflecting socio-cultural backgrounds. Perceptions influence individual and collective preparedness, response and recovery to short term extreme weather events, as well as people's adaptive behaviours to long term change, such as sea-level rise. Experiential and socio-cultural factors may explain significantly more variance in climate risk perception than either cognitive or socio-demographic characteristics (van der Linden 2015). Previous experiences of loss and damage can also shape expectations about the prevalence and severity of future events such that perception of risk increases sharply after exposure to flooding (e.g. Botzen et al. 2009; Kellens et al. 2012; Gallagher 2014) and makes people more willing to make household level changes and be better prepared (Lawrence et al. 2014). Even within the same household, climate risk perception and adaptive responses differ between genders for the same shock (Mishra and Pede 2017). Individuals may change their perception of risk over time either as the result of direct experience of one or more hazards or based on new information acquired through trusted social networks or other information sources (e.g. Magliocca and Walls 2018).

The challenge in deltas in poorer countries is to address the cyclical and chronic changes in the deltaic environment, the frequent hazards, poverty and the need for economic development, alongside the increasing ad hoc physical modifications of canals, dykes and polders. For example, in the Mahanadi Delta in India, repeated cycles of disasters coupled with recurrent (and expensive) cultural activities, ineffective livelihood diversification, ineffective formal institutional support and limited access to land all combine to reduce individual and household resilience to hazards (Duncan et al. 2017). The solutions used to address past and present challenges could change the future for delta residents; the following sections address the questions how might that happen? What policies are used in deltas? And, what might future transformational adaptation policies for deltas look like?

9.3 Adaptation Policies and Incentives in Deltas

Many elements of policies for adaptation to environmental risks in deltas mirror planning and policymaking in other low lying coastal areas: policy options are largely described within the broad concepts of protect, accommodate or retreat (Bijlsma et al. 1996), also referred to as armour, adapt or retreat. Deltas are also widely referred to as poverty, climatic and development hotspots (de Souza et al. 2015). Specific delta policies or strategies are largely sets of principles framed around a broad geographical area. They include the Dutch Delta Programme, the 2016–2019 Mississippi Delta Region Development Plan, the Niger Delta Master Plan, the Mekong Delta Plan, or the Bangladesh Delta Plan 2100 (Seijger et al. 2017). However, there are not comprehensive delta plans or processes for many of the world's most significant and populous deltas (see Chapters 2–4, Mensah et al. [2016] and Hazra et al. [2016]).

Policy choices for deltas have been influenced by international cooperation and treaties such as the Sendai Framework for disaster risk reduction agenda, UNFCCC climate change adaptation reporting requirements, and the Sustainable Development Goals (Lwasa 2015). At the subnational scale, adaptation policy appears to be largely focussed on addressing disaster risk, yet there is only limited documentation of the initiatives that are taking place e.g. managing coastal erosion through creation of barriers, storm surge barriers, adaptation of housing to flooding (Kates et al. 2012). Of all the adaptation policies in deltas, these can be grouped into three main components: addressing pre-existing socio-economic vulnerability, reducing disaster risk and building long term social-ecological resilience, see Fig. 9.1.

In many deltas, much of the current effort in adaptation policy is focussed on reducing vulnerability. For example, Bangladesh formulated the Climate Change Strategy and Action Plan 2009 (MoEF 2009) and established a climate change trust fund in 2010 to fund implementation (Ayers et al. 2014). The strategy and action plan proposed six areas of activity namely: food security, social protection and health; comprehensive disaster management; infrastructure; research and knowledge management; mitigation and low carbon development; and capacity building and institutional strengthening (Islam and Nursey-Bray 2017). Within these areas, 44 programmes have been funded to date, and are categorised in Fig. 9.2 to be distributed among three elements: vulnerability reduction, disaster risk reduction and ecological resilience (Tompkins et al. 2018).

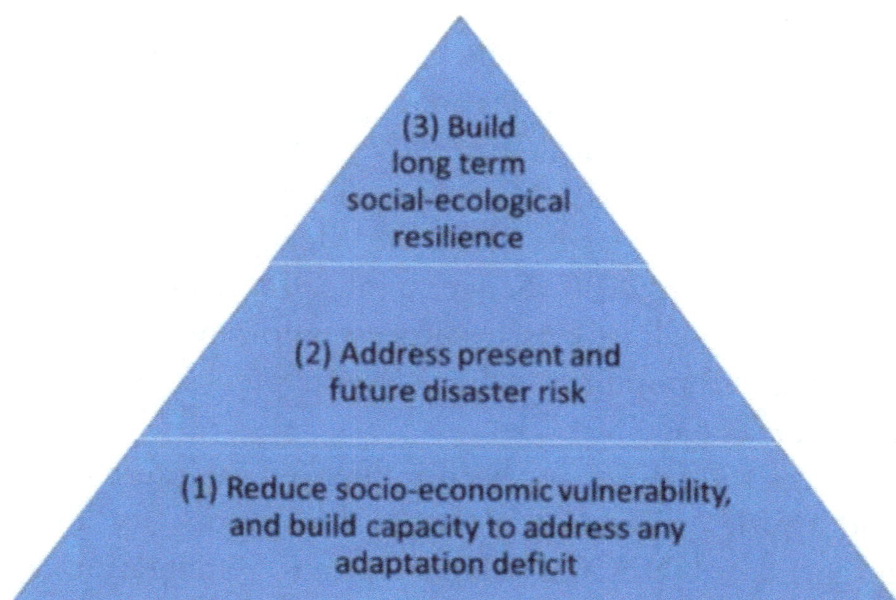

Fig. 9.1 Components of adaptation policy in deltas (Adapted from Tompkins et al. [2018] under CC BY 4.0)

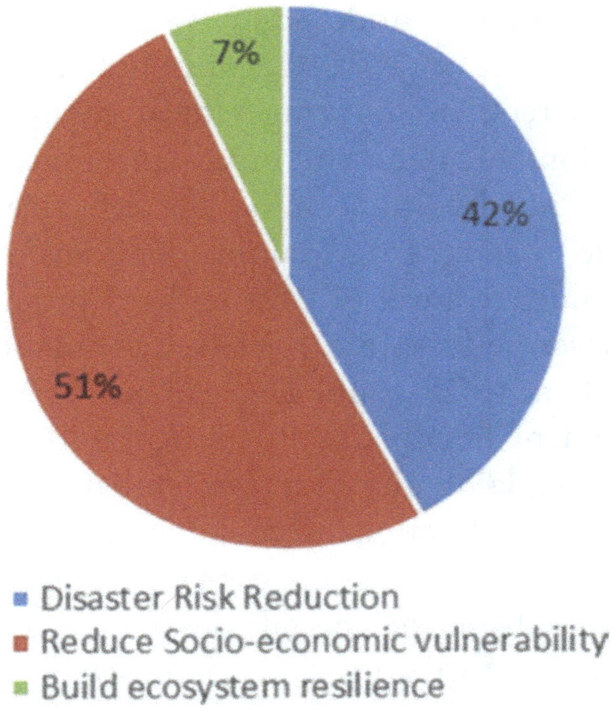

Fig. 9.2 Distribution of types of adaptations across the GBM Delta undertaken by Bangladesh Climate Change Trust (BCCTF) during 2009–2017 (Data from: Annual Reports since 2009 of Bangladesh Climate Change Trust, Ministry of Environment and Forests, Dhaka. Adaptation types follow Tompkins et al. [2017])

Measures in the delta include construction/repair of embankments, river bank protection, cyclone shelters, etc. The vulnerability reducing measures include re-excavation of canals, improving drinking water supply, raising homesteads, etc. Ecosystem based adaptations include coastal mangrove plantation. Of the 231 measures considered, about 80% of the total investment has been made in food security and infrastructure clusters. Very little investment has been made in research and capacity building. Among various ministries in Bangladesh, the Ministry of Water Resources, Ministry of Local Government and Ministry of Environment and Forest received most funds. Local government institutions received much less funding compared to central agencies, but performed better in targeting adaptation deficits and mainstreaming gender considerations (Vij et al. 2018).

Common adaptation policies and programmes that seek to reduce vulnerability are typically incremental (Denton et al. 2014).

Beyond the developing world, some policies seek to be more transformational (Kates et al. 2012) by fundamentally changing the nature of a system, or inducing radical change across systems. Such transformations focus on the future and long term substantial change, and may involve questioning the effectiveness of existing systems (Lonsdale et al. 2015). Examples of transformational adaptation policy include: removal of existing hard protection and barriers to tidal and riverine flow (e.g. riverine and tidal dike removal) in the Mississippi Delta (Mississippi Department of Marine Resources 2011); reactivation of floodplains in the Rhine delta (ICPR 2015); and restoring floodplains that remove embankments and return agricultural polders to floodplains to increase floodwater retention capacity in the Yangtze (Chen et al. 2014). Managed retreat of infrastructure and people from the coastal Mekong (USAID 2014) represents a significant transformation and demonstrates that such radical plans often have significant losers as well as winners. All of these policy choices reveal a dramatic shift away from current and historical adaptation policy choices in the various deltas (Vincent 2017). It is in this context that individual households, businesses and communities are adapting to shocks and stresses. The following sections consider: how are people adapting and how is policy affecting adaptation choices? What adaptations are considered effective and is there agreement on the best way to adapt?

9.4 Adapting to Present Day Stresses

Despite a long history of adaptation to environmental change in deltas, little is known about the specifics of this adaptation, for example, who is adapting, how and why, and how this has changed over time. However, given the ambition to undertake a global stocktake of adaptation by 2023 as mandated by the Paris Agreement to the UNFCCC, documentation of such adaptation practice is urgently required (Tompkins et al. 2018). At present, it is known that households and individuals do not adapt in isolation from the national policy context, but operate within it. Household choices are mediated by a number of factors, including non-government organisations (NGOs), international advocacy groups, the private sector and the socio-cultural context. Within the current research, drawing on multi-scale governance literatures, a typology of the factors influencing how policy and household choices interact is identified (Fig. 9.3).

Adaptation policy can play a role in supporting adaptation. For example, support to convert land to alternative livelihoods, such as horticulture, or resourcing to support community-based cyclone preparedness activity, can spur on households to undertake adaptive actions. However, the extent to which policies achieve their intended goals is variable. Cyclone shelters are installed to provide shelter during and after extreme events, yet women and girls are often reluctant to stay in public shelters where they may have to interact with men, to maintain honour and avoid shame and harassment (Rashid and Michaud 2002; Juran and Trivedi 2015). Poverty can constrain household adaptation choice. For example, government policy in India provides training for farmers on climate tolerant crop varieties to improve agricultural productivity in increasingly saline or dry conditions. However, poor farmers may not have the time to travel to training on new crop varieties, or have the buffering capacity to take the chance to change crops just in case of crop failure.

While many adaptation policies have been put in place, imperfect implementation can also mean that the social consequences have not always been even (Mimura et al. 2014). In Bangladesh, dykes and polders are essential to protect properties and agricultural fields from tidal

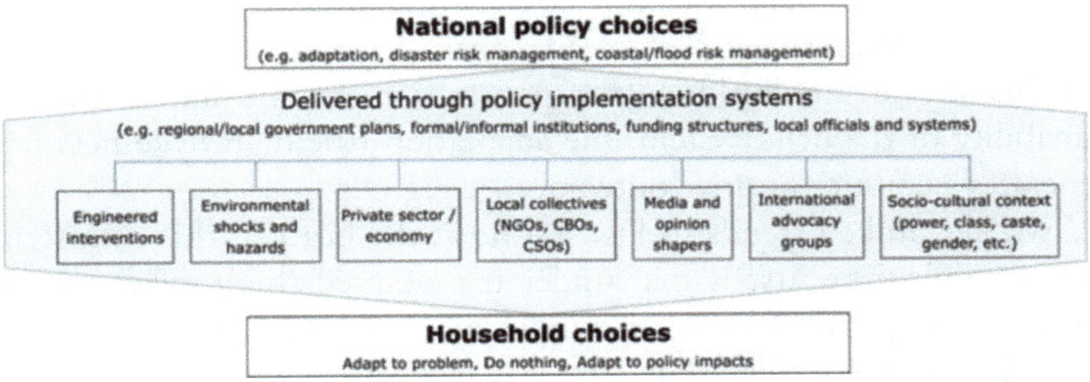

Fig. 9.3 National policy influences adaptation choices by households, mediated by social and environmental factors

flooding. Many of these polders are still awaiting rehabilitation following severe cyclone damage in 2007 and 2009. This has prolonged community suffering due to the continued threat of tidal flooding, income insecurity, lack of freshwater supply and ongoing vulnerabilities due to weak coastal embankments remain a concern long after those cyclones (Sadik et al. 2018). In the Mekong a government programme of dyke building has enabled multiple crops per year, encouraging commercial production and leading to a reduction in small-scale farming and net out-migration of people from the delta (Chapman et al. 2016). As highlighted in Chapter 7, government action is patchy: not all communities that require relocation, or demand it, are necessarily included in plans (Mortreux et al. 2018).

Adaptation policy choices can also lead to unexpected impacts, where individuals have to adapt to the consequences of the adaptation. In the Vietnamese Mekong Delta, the most profound recent effort has been the creation of an extensive high dyke network, spanning thousands of kilometres and encompassing the majority of the delta's rice paddies. Much of the effort in creating this network occurred during the late 1990s and early 2000s. Through household survey (Chapman et al. 2016), creation of a system dynamics model (Chapman 2016) and multi-criteria analysis (Chapman 2016), findings suggest that extending and heightening the Vietnamese Mekong Delta dyke network is an effective adaptation against prevalence of extreme river flood events. However, this finding is only true when greater weight is placed on large-scale short-term food production and export, and the incomes of wealthier (large land-owning) farmers. Should decision-makers take a pro-poor approach, and place an equal or greater weighting on the sustainability of the livelihoods of poorer farmers, and indeed the sustainability of the delta system, the adaptation (the high dyke network) generates a counterintuitive outcome, see Fig. 9.4.

Two key, linked trends in Fig. 9.4 lie at the heart of this counterintuitive result. The first is that under the adapted (high dyke) system, rates of change over time in rice input efficiencies (i.e. yield per tonne of fertiliser) reverse direction. The loss of nutrient-rich sediment deposition (in the unadapted system), historically brought by the now excluded flood, degrades the quality of the soil and pushes farmers

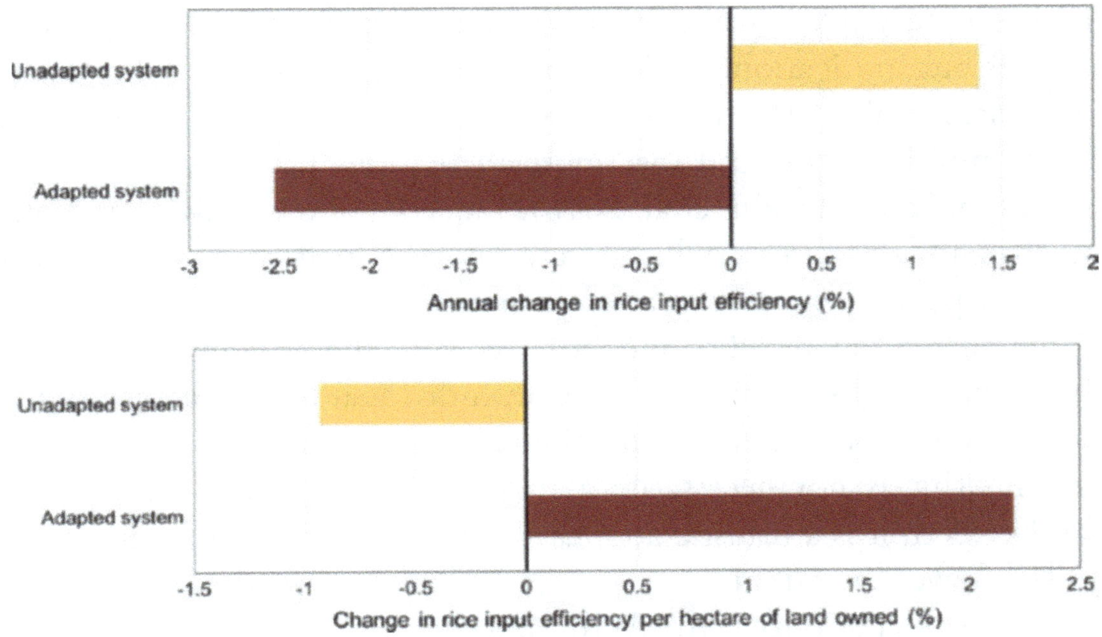

Fig. 9.4 The distributional impacts of adaptation in rice farming systems in the Vietnamese Mekong Delta (Adapted from Chapman et al. [2016] under CC BY 4.0)

towards heavier fertiliser use. The second trend, a direct result of this, is a reversal in the relative advantage of farm size (Fig. 9.4), from favouring smaller operations (unadapted system), to favouring larger operations (adapted systems). Small-scale operations of one hectare or less tend to lack the resources to compete in a fertiliser-intensive system, having previously benefitted from the free provision of sediment-bound nutrients. Chapman et al. (2016) point to the importance of recognising whose priorities count in evaluating the success of adaptation policy choices. They acknowledge that the success of adaptations is normative: there are winners and losers, and trade-offs will always be needed (see also Hutton et al. 2018).

A key issue raised in policy and science is the assessment of success in adaptation. Under what conditions can adaptation be considered a success? And how does success vary with social factors, such as gender, age and caste? There are various criteria by which adaptation can be evaluated, for example effectiveness in terms of long-term sustainable development, cost-efficiency of the action, equity of the distribution of impacts or the legitimacy of the action (Adger et al. 2005).

Despite some work considering no-regrets adaptation and adaptations that generate mitigation or developmental co-benefits (see, for example, Suckall et al. 2015), it is broadly agreed that there is little evidence of such multiple wins and that most adaptations have negative consequences for some (Ficklin et al. 2018). Indeed, there is growing recognition that no adaptations will generate universal benefits, and there will always be people who lose as a result of adaptation, either through paying for adaptation benefits and not receiving them, being affected by others' adaptations, or even because an individual has no choice but to adapt in a way that does not contribute to long-term sustainable development. This is not necessarily a message that policymakers wish to hear, however, it is a realistic appraisal of the impacts of adaptation policy. This leaves the question—how can inclusive adaptation strategies be designed for deltas during the Anthropocene?

9.5 The Design of Inclusive Adaptation Strategies

Adaptation is a spatially and temporally dynamic process with accrued benefits potentially changing with geography, time and circumstances. What might be considered an effective adaptation response in one place at one time may, with time, become less or more effective with associated consequences and potentially bringing into question the sustainability of the adaptive response. An example is the Mekong Delta in Vietnam where short-term benefits of engineering interventions to increase rice production from two annual crops to three are offset by the longer term impacts. The impacts include: soil quality degradation associated with fertiliser use, reduction in fishery co-production and loss of ecosystem services from the introduction of agricultural pest predators associated with flooding (Chapman et al. 2016). In Bangladesh, there can be long term financial benefits of enhanced horticulture production, in lieu of traditional rice farming, but due to the highly variable year on year yields of horticulture (which can create lean years), there is much lower uptake of this adaptation by poorer socio-ecological groups on the delta (Hutton et al. 2018).

Evidence from past adaptations in deltas reveals a spectrum of initiatives, from extensive investments in cyclone preparedness and recovery in Bangladesh (Mallick and Rahman 2013), to dyke and polder building in India, and construction of embankments in the United States. Other significant change has occurred in deltas as a result of social policy. For example, the Mahatma Gandhi Rural Employment Guarantee Scheme in India provides a social safety net for those who are below the poverty line. There is also evidence of attempts to redraw, what often tend to be, entrenched patterns of competition and dominance in the allocation of water and management of river basins (Budds 2013). Adaptation strategies are clearly not simply engineered solutions, nor are they simply social policies, they are a complex web of policies that affect the various components within deltas: land use, ecosystems, rural and urban development, transport, disaster risk reduction, to name a few, see, for example, Mensah et al. (2016), Hazra et al. (2016), and Dey et al. (2016).

Recent research has endeavoured to consider what future strategies in deltas may look like (Suckall et al. 2015). Adaptation is not limited to one sector, but needs to be considered in the light of the bigger picture. Policymakers are often lacking insights into how policy can affect adaptation strategies, and the trade-offs that need to be made to reflect normative goals (for example equitable poverty reduction, or emphasis on national level economic growth). In turn, adaptation policy choices are affected by the costs of adaptation and the extent to which policy change, and political effort, is required.

Suckall et al. (2018) develop narratives of adaptation policy—which comprise multiple policies in the areas of: addressing disaster risk, reducing socio-economic vulnerability and managing landscapes and ecological systems. Each adaptation policy direction requires different levels of investment, resourcing and policy support. *Minimum intervention* brings together policy choices that could be explained as focusing on low cost adaptation policies designed to achieve maximum impact. The focus here tends to on basic emergency response to disasters. *Capacity expansion* encourages climate-proof economic growth, requiring investment, but does not seek to make significant change to the current structure of the economy. *Efficiency enhancement* requires less investment than policy commitment, is an ambitious strategy that promotes adaptation consistent with the most efficient management and exploitation of the current system, looking at ways of distributing labour, balancing livelihood choices and best utilising ecosystem services to enhance livelihoods and wellbeing under climate change. *System restructuring* requires the greatest level of investment and policy commitment, and is based on pre-emptive fundamental change at every level in order to completely transform the current social and ecological system, and change the social and physical functioning of the delta system. This argues that the system can be restructured in one of three main ways—each with a different focus: protection, accommodating change and retreating/moving away. Each has a different end goal for the delta. Collectively these policy directions offer insight to policymakers by envisioning what policy direction opportunities there are for deltaic regions. Under a changing climate, with an inevitable reduction in sediment in deltas resulting from upstream damming and other land use modifications (see Chapter 6), it is known that the structure of deltas will substantially, and potentially fundamentally, change further in the Anthropocene. What are the implications of these different adaptation policy directions in this context? Chapter 10 considers how these different policy directions can generate different outcomes for deltas, and it considers explicitly the trade-offs that need to be made to achieve policy goals.

9.6 Conclusions

Policies and planning will play a powerful role in creating the human-dominated deltas of the Anthropocene. There is good evidence that active management of deltas can potentially generate sustainability for deltas and their populations. Governments retain the autonomy to identify their priorities for development of many deltas, and choose adaptation policy directions that help to achieve these aims.

The emergence of delta management plans in many developed and developing countries is a positive sign of proactive attempts to manage the complex interactions of natural environments and human systems in the Anthropocene. A key insight is that what deltas will look like in the future depends substantially on current policy directions and choices. This is not a simple policy choice particularly due to gaps in our information and understanding about what adaptations are most successful over time, and across which population groups. The global stocktake of adaptation mandated by the Paris Agreement of the UNFCCC will provide insights into the prevalence and quality of adaptations, including in deltas, to understand who is gaining and who is losing from alternative adaptation policy directions.

National governments with deltas within their boundaries, and neighbouring countries in watersheds are facing major challenges in managing deltas in a changing climate, not least with limited information about directions and motivations for adaptation of diverse actors. Key dilemmas for governments in delta adaptation policy directions are to decide whose voice should be heard in developing plans, and what trade-offs they would consider to be unacceptable. In this way, planning for an uncertain future can proceed on a sound basis.

References

Adger, W. N., Arnell, N. W., & Tompkins, E. L. (2005). Successful adaptation to climate change across scales. *Global Environmental Change, 15*(2), 77–86.

Ayers, J. M., Huq, S., Faisal, A. M., & Hussain, S. T. (2014). Mainstreaming climate change adaptation into development: A case study of Bangladesh. *Wiley Interdisciplinary Reviews: Climate Change, 5*(1), 37–51. https://doi.org/10.1002/wcc.226.

Bijlsma, L., Ehler, C. N., Klein, R. J. T., Kulshrestha, S. M., McLean, R. F., Nimura, N., et al.. (1996). Coastal zones and small islands. In R. T. Watson, M. C. Zinyowera, & R. H. Moss (Eds.), *Climate change 1995: Impacts adaptations and mitigation of climate change: Scientific-technical analysis* (pp. 293–342). Contribution of Working Group II to the Second Assessment Report of the Intergovermental Panel on Climate Change. Cambridge: Cambridge University Press.

Botzen, W. J. W., Aerts, J. C. J. H., & van den Bergh, J. C. J. M. (2009). Dependence of flood risk perceptions on socioeconomic and objective risk factors. *Water Resources Research, 45,* 10. https://doi.org/10.1029/2009WR007743.

Budds, J. (2013). Water, power, and the production of neoliberalism in Chile, 1973–2005. *Environment and Planning D: Society and Space, 31*(2), 301–318. https://doi.org/10.1068/d9511.

Chapman, A. (2016). *Evaluating an adaptation: Rice-sediment trade-offs in the Vietnamese Mekong Delta* (PhD thesis). Faculty of Social, Human and Mathematical Sciences, University of Southampton. https://eprints.soton.ac.uk/400969/.

Chapman, A. D., Darby, S. E., Hồng, H. M., Tompkins, E. L., & Van, T. P. D. (2016). Adaptation and development trade-offs: Fluvial sediment deposition and the sustainability of rice-cropping in An Giang Province, Mekong Delta. *Climatic Change*, 1–16. http://dx.doi.org/10.1007/s10584-016-1684-3.

Chen, Y., Yu, X., Pittock, J., & Jiang, L. (Eds.). (2014). *Ecosystem services and management strategy in China.* Springer Earth System Sciences. Berlin: Springer.

Denton, F., Wilbanks, T. J., Abeysinghe, A. C., Burton, I., Gao, Q., Lemos, M. C., et al. (2014). Climate-resilient pathways: Adaptation, mitigation, and sustainable development. In C. B. Field, V. R. Barros, D. J. Dokken, K. J. Mach, M. D. Mastrandrea, T. E. Bilir, et al. (Eds.), *Climate change 2014: Impacts, adaptation, and vulnerability: Part A: Global and sectoral aspects* (pp. 1101–1131). Contribution of Working Group II to the Fifth Assessment Report of the Intergovernmental Panel on Climate Change. Cambridge UK and New York, NY: Cambridge University Press.

de Souza, K., Kituyi, E., Harvey, B., Leone, M., Murali, K. S., & Ford, J. D. (2015). Vulnerability to climate change in three hot spots in Africa and Asia: Key issues for policy-relevant adaptation and resilience-building research. *Regional Environmental Change, 15*(5), 747–753. https://doi.org/10.1007/s10113-015-0755-8.

Dey, S., Ghosh, A. K., & Hazra, S. (2016). *Review of West Bengal State adaptation policies, Indian Bengal Delta* (Deltas, Vulnerability and Climate Change: Migration and Adaptation [DECCMA] Working Paper). Southampton, UK: DECCMA Consortium. https://generic.wordpress.soton.ac.uk/deccma/resources/working-papers/. Last accessed 6 August 2018.

Duncan, J. M., Tompkins, E. L., Dash, J., & Tripathy, B. (2017). Resilience to hazards: Rice farmers in the Mahanadi Delta. *India. Ecology and Society, 22,* 4. https://doi.org/10.5751/ES-09559-220403.

Fergusson, J. (1863). On recent changes in the delta of the Ganges. *Quarterly Journal of the Geological Society, 19*(1–2), 321–354.

Ficklin, L., Stringer, L. C., Dougill, A. J., & Sallu, S. M. (2018). Climate compatible development reconsidered: Calling for a critical perspective. *Climate and Development, 10*(3), 193–196. https://doi.org/10.1080/17565529.2017.1372260.

Gallagher, J. (2014). Learning about an infrequent event: Evidence from flood insurance take-up in the United States. *American Economic Journal: Applied Economics, 6*(3), 206–233.

Hazra, S., Dey, S., & Ghosh, A. K. (2016). *Review of Odisha State adaptation policies, Mahanadi Delta* (Deltas, Vulnerability and Climate Change: Migration and Adaptation [DECCMA] Working Paper). Southampton, UK: DECCMA Consortium. https://generic.wordpress.soton.ac.uk/deccma/resources/working-papers/. Last accessed 6 August 2018.

Hutton, C. W., Nicholls, R. J., Lázár, A. N., Chapman, A., Schaafsma, M., & Salehin, M. (2018). Potential trade-offs between the Sustainable Development Goals in Coastal Bangladesh. *Sustainability, 10*(4), 1008. http://dx.doi.org/10.3390/su10041108.

ICPR. (2015). *Strategy for the International River Basin District (IRBD) Rhine for adapting to climate change* (Report No. 219). Koblenz, Germany: International Commission for the Protection of the Rhine (ICPR). https://www.iksr.org/fileadmin/user_upload/DKDM/Dokumente/Fachberichte/EN/rp_En_0219.pdf. Last accessed 22 August 2018.

Islam, M. T., & Nursey-Bray, M. (2017). Adaptation to climate change in agriculture in Bangladesh: The role of formal institutions. *Journal of Environmental Management, 200,* 347–358. https://doi.org/10.1016/j.jenvman.2017.05.092.

Juran, L., & Trivedi, J. (2015). Women, gender norms, and natural disasters in Bangladesh. *Geographical Review, 105*(4), 601–611. https://doi.org/10.1111/j.1931-0846.2015.12089.x.

Kaijser, A., & Kronsell, A. (2014). Climate change through the lens of intersectionality. *Environmental Politics, 23*(3), 417–433. https://doi.org/10.1080/09644016.2013.835203.

Kates, R. W., Travis, W. R., & Wilbanks, T. J. (2012). Transformational adaptation when incremental adaptations to climate change are insufficient. *Proceedings of the National Academy of Sciences, 109,* 7156–7161.

Kellens, W., Terpstra, T., & De Maeyer, P. (2012). Perception and communication of flood risks: A systematic review of empirical research. *Risk Analysis, 33*(1), 24–49. https://doi.org/10.1111/j.1539-6924.2012.01844.x.

Lawrence, J., Quade, D., & Becker, J. (2014). Integrating the effects of flood experience on risk perception with responses to changing climate risk. *Natural Hazards, 74*(3), 1773–1794. https://doi.org/10.1007/s11069-014-1288-z.

Ling, F. H., Tamura, M., Yasuhara, K., Ajima, K., & Trinh, C. V. (2015). Reducing flood risks in rural households: Survey of perception and adaptation in the Mekong Delta. *Climatic Change, 132*(2), 209–222. https://doi.org/10.1007/s10584-015-1416-0.

Lonsdale, K., Pringle, P., & Turner, B. (2015). *Transformative adaptation: What it is, why it matters and what is needed.* UK Climate Impacts Programme. University of Oxford, Oxford, UK. https://ukcip.ouce.ox.ac.uk/wp-content/PDFs/UKCIP-transformational-adaptation-final.pdf. Last accessed 20 August 2018.

Lwasa, S. (2015). A systematic review of research on climate change adaptation policy and practice in Africa and South Asia deltas. *Regional Environmental Change, 15*(5), 815–824.

Magliocca, N. R., & Walls, M. (2018). The role of subjective risk perceptions in shaping coastal development dynamics. *Computers, Environment and Urban Systems, 71,* 1–13. https://doi.org/10.1016/j.compenvurbsys.2018.03.009.

Mallick, F., & Rahman, A. (2013). Cyclone and tornado risk and reduction approaches in Bangladesh. In R. Shaw, F. Mallick, & A. Islam (Eds.), *Disaster risk reduction approaches in Bangladesh* (pp. 91–102). Tokyo, Japan: Springer. http://dx.doi.org/10.1007/978-4-431-54252-0_5.

Mensah, A., Anderson, K., & Nelson, W. (2016). *Review of adaptation related policies in Ghana* (Deltas, Vulnerability and Climate Change: Migration and Adaptation [DECCMA] Working Paper). Southampton, UK: International Development Research Centre (IDRC). http://www.deccma.com/deccma/Working_Papers/. Last accessed 20 August 2018.

Mimura, N., Pulwarty, R. S., Duc, D. M., Elshinnawy, I., Redsteer, M. H., Huang, H. Q., et al. (2014). Adaptation planning and implementation. In C. B. Field, V. R. Barros, D. J. Dokken, K. J. Mach, M. D. Mastrandrea, T. E. Bilir, et al. (Eds.), *Climate change 2014: Impacts, adaptation, and vulnerability: Part A: Global and sectoral aspects* (pp. 869–898). Contribution of Working Group II to the Fifth Assessment Report of the Intergovernmental Panel on Climate Change. Cambridge, UK and New York, NY: Cambridge University Press.

Mishra, A. K., & Pede, V. O. (2017). Perception of climate change and adaptation strategies in Vietnam: Are there intra-household gender differences? *International Journal of Climate Change Strategies and Management, 9*(4), 501–516. https://doi.org/10.1108/IJCCSM-01-2017-0014.

Mississippi Department of Marine Resources. (2011). *Assessment of sea level rise in coastal Mississippi* (Report submitted to the Office of Coastal management and Planning by Eco-System Inc). Biloxi, MS: Mississippi Department of Marine Resources. http://www.dmr.state.ms.us/images/cmp/2011-slr-final.pdf. Last accessed 15 August 2018.

MoEF. (2009). *Bangladesh climate change strategy and action plan 2009* (Dhaka, Bangladesh, Ministry of Environment and Forestry [MoEF]). Government of the People's Republic of Bangladesh. https://www.iucn.org/content/bangladesh-climate-change-strategy-and-action-plan-2009. Last accessed 8 October 2018.

Mortreux, C., Safra de Campos, R., Adger, W. N., Ghosh, T., Das, S., Adams, H., et al. (2018). Political economy of planned relocation: A model of action and inaction in government responses. *Global Environmental Change, 50,* 123–132. https://doi.org/10.1016/j.gloenvcha.2018.03.008.

Nicholls, R. J., Wong, P. P., Burkett, V. R., Codignotto, J. O., Hay, J. E., McLean, R. F., et al. (2007). Coastal systems and low-lying areas. In M. L. Parry, O. F. Canziani, J. P. Palutikof, P. J. van der Linden, & C. E. Hanson (Eds.), *Climate change 2007: Impacts, adaptation and vulnerability* (pp. 315–356). Contribution of Working Group II to the Fourth Assessment Report of the Intergovernmental Panel on Climate Change. Cambridge, UK: Cambridge University Press.

Pearse, R. (2016). Gender and climate change. *Wiley Interdisciplinary Reviews: Climate Change, 8*(2), e451. http://dx.doi.org/10.1002/wcc.451.

RAJUK. (2015). Chapter 2: Dhaka past and present. *Dhaka Structure Plan 2016–2035*. Dhaka, Bangladesh: Rajdhani Unnayan Kartripakkha (RAJUK). http://www.rajukdhaka.gov.bd/rajuk/image/slideshow/5.Chapter%2002.pdf. Last accessed 14 August 2018.

Rashid, S. F., & Michaud, S. (2002). Female adolescents and their sexuality: Notions of honour, shame, purity and pollution during the floods. *Disasters, 24*(1), 54–70. https://doi.org/10.1111/1467-7717.00131.

Ray-Bennett, N. S. (2009). The influence of caste, class and gender in surviving multiple disasters: A case study from Orissa, India. *Environmental Hazards, 8*(1), 5–22. https://doi.org/10.3763/ehaz.2009.0001.

Sadik, M. S., Nakagawa, H., Rahman, R., Shaw, R., Kawaike, K., & Fujita, K. (2018). A study on Cyclone Aila recovery in Koyra, Bangladesh: Evaluating the inclusiveness of recovery with respect to predisaster vulnerability reduction. *International Journal of Disaster Risk Science, 9*(1), 28–43. https://doi.org/10.1007/s13753-018-0166-9.

Seijger, C., Douven, W., van Halsema, G., Hermans, L., Evers, J., Phi, H. L., et al. (2017). An analytical framework for strategic delta planning: Negotiating consent for long-term sustainable delta development. *Journal of Environmental Planning and Management, 60*(8), 1485–1509. https://doi.org/10.1080/09640568.2016.1231667.

Suckall, N., Stringer, L. C., & Tompkins, E. L. (2015). Presenting triple-wins? Assessing projects that deliver adaptation, mitigation and development co-benefits in rural Sub-Saharan Africa. *Ambio, 44*(1), 34–41. https://doi.org/10.1007/s13280-014-0520-0.

Suckall, N., Tompkins, E. L., Nicholls, R. J., Kebede, A. S., Lázár, A. N., Hutton, C., et al. (2018). A framework for identifying and selecting long term adaptation policy directions for deltas. *Science of the Total Environment, 633*, 946–957.

Szabo, S., Brondizio, E., Renaud, F. G., Hetrick, S., Nicholls, R. J., Matthews, Z., et al. (2016). Population dynamics, delta vulnerability and environmental change: Comparison of the Mekong, Ganges-Brahmaputra and Amazon delta regions. *Sustainability Science, 11*(4), 539–554. https://doi.org/10.1007/s11625-016-0372-6.

Tessler, Z., Vörösmarty, C. J., Grossberg, M., Gladkova, I., Aizenman, H., Syvitski, J., et al. (2015). Profiling risk and sustainability in coastal deltas of the world. *Science, 349*(6248), 638–643. https://doi.org/10.1126/science.aab3574.

Tompkins, E. L., Suckall, N., Vincent, K., Rahman, R., Mensah, A., Ghosh, A., et al. (2017). *Observed adaptation in deltas* (Deltas, Vulnerability and Climate Change: Migration and Adaptation [DECCMA] Working Paper). Southampton, UK: University of Southampton. http://www.deccma.com/deccma/Working_Papers/. Last accessed 7 November 2018.

Tompkins, E. L., Vincent, K., Nicholls, R. J., & Suckall, N. (2018). Documenting the state of adaptation for the global stocktake of the Paris Agreement. *WIREs Climate Change*. http://dx.doi.org/10.1002/wcc.545.

USAID. (2014). *USAID Mekong ARCC climate change impact and adaptation study for the Lower Mekong Basin* (Summary report). Prepared by the International Centre for Environmental Management. United States Agency for International Development (USAID), Bangkok, Thailand. https://dec.usaid.gov/dec/home/Default.aspx. Last accessed 28 August 2018.

van der Linden, S. (2015). The social-psychological determinants of climate change risk perceptions: Towards a comprehensive model. *Journal of Environmental Psychology, 41,* 112–124. https://doi.org/10.1016/j.jenvp.2014.11.012.

Vermaat, J. E., & Eleveld, M. A. (2013). Divergent options to cope with vulnerability in subsiding deltas. *Climatic Change, 117*(1–2), 31–39.

Vij, S., Biesbroek, R., Groot, A., & Termeer, K. (2018). Changing climate policy paradigms in Bangladesh and Nepal. *Environmental Science & Policy, 81,* 77–85. https://doi.org/10.1016/j.envsci.2017.12.010.

Vincent, K. (2017). *Transformational adaptation: A review of examples from 4 deltas to inform the design of DECCMA's adaptation policy trajectories* (Deltas, Vulnerability and Climate Change: Migration and Adaptation [DECCMA] Working Paper) (p. 18). Southampton, UK: DECCMA Consortium. https://generic.wordpress.soton.ac.uk/deccma/resources/working-papers/. Last accessed 20 August 2018.

Wolters, M. L., Sun, Z., Huang, C., & Kuenzer, C. (2016). Environmental awareness and vulnerability in the Yellow River Delta: Results based on a comprehensive household survey. *Ocean and Coastal Management, 120,* 1–10. https://doi.org/10.1016/j.ocecoaman.2015.11.009.

10

Choices: Future Trade-Offs and Plausible Pathways

Attila N. Lázár, Susan E. Hanson, Robert J. Nicholls, Andrew Allan, Craig W. Hutton, Mashfiqus Salehin and Abiy S. Kebede

10.1 Introduction

Human-dominated deltas are hotspots of high population density, economic development and ecosystem service value that are concurrently subject to natural and anthropogenic pressures and changes. The integrity of deltas, as coupled human and natural systems, is fundamentally

A. N. Lázár (✉) · S. E. Hanson · R. J. Nicholls · A. S. Kebede
School of Engineering, University of Southampton, Southampton, UK
e-mail: a.lazar@soton.ac.uk

A. Allan
School of Law, University of Dundee, Dundee, UK

C. W. Hutton
GeoData Institute, Geography and Environmental Science, University of Southampton, Southampton, UK

M. Salehin
Institute of Water and Flood Management, Bangladesh University of Engineering and Technology, Dhaka, Bangladesh

R. J. Nicholls et al. (eds.), *Deltas in the Anthropocene*,
https://doi.org/10.1007/978-3-030-23517-8_10

integrated with their hydrodynamic processes and in particular the provision of sediments (see Chapters 5 and 6). A natural delta adjusts its form and function in balance with such processes (Wolman and Gerson 1978), but Anthropocene deltas are more defined by the intimate relationship between humans and their physical systems (see Day et al. 2016). Human settlements and economies manipulate the natural circumstances in the delta for their benefit, but when such efforts occur on the catchment scale, this permanently alters the trends and balance of the whole system (Overeem and Syvitski 2009).

Contemporary trends and changes inevitably lead to trade-offs between the needs of diverse populations of people and the requirements to maintain a sustainable, productive environment and associated services (Steffan-Dewenter et al. 2007; Hutton et al. 2018). For example, to accept annual floods would be beneficial for maintaining long-term land elevation, soil fertility and local ecosystems (Chapter 5). However, benefits of incorporating natural processes into already engineered delta systems come with the cost of potential flood damage to infrastructures, people and agriculture. Other trade-offs are between development and nature at various scales. Land transformations to aquaculture in brackish conditions, for example, results in soil and biodiversity degradation and decline in local agriculture balanced against economic needs at a national level (Amoako-Johnson et al. 2016). However, populations are not static, adapting and evolving temporally and spatially within the Anthropocene (as discussed in Chapters 7 and 9). Indeed, individual households and communities adapt to changing circumstances, in effect making trade-offs at community and household level. Both natural resource-based and off-farm households constantly adapt due to a range of reasons including environmental pressures and economic opportunities.

Policies aiming to promote particular human benefits can also often generate diverse and unforeseen trade-offs at the government, community or individual level, and over short- or long-term timescales. Focussing on identifying, understanding and quantifying such trade-offs enables governments, planners and civil society to accomplish their national aims while supporting delta populations in their development.

In this chapter therefore, choices, potential trade-offs and plausible pathways for Anthropocene deltas are explored, drawing fundamentally on integrated modelling and other simulation approaches. Trade-offs from coastal Bangladesh in the Ganges-Brahmaputra-Meghna (GBM) Delta in particular are outlined to illustrate environmental and socio-economic trends, potential trade-offs and local adaptation options as a result of high-level policy decisions. The analysis includes actions and dilemmas faced by individuals, referred to as household adaptation (see Chapter 9) across the three deltas considered in Chapters 2–4. This analysis is used to reflect on delta management across the world and how analysis of trade-offs can be used to address future development trajectories.

10.2 Policy Development in Deltas

A trade-off is "*a situation where one balances two opposing situations or qualities*", which inherently means that "*one accepts something bad in order to have something good*" (Online Cambridge Dictionary). Understanding trade-offs is especially crucial for deltas as their entire extent and existence are uniquely vulnerable to decisions made at national, local and household levels. Broader governance frameworks across a range of scales need to consider exogenous and endogenous factors over the delta and its catchment in determining how multiple interests can be weighed over time and space.

Governance frameworks and regimes serve at least in part to facilitate the balancing the needs of sectors, interests and regions (Paavola et al. 2009). Higher quality systems are able to maximise the accountability of those making and implementing decisions and hopefully ensure that, while decisions will inevitably be taken in the absence of perfect information, appropriate adaptation is possible. Coordination and integration of policymaking across sectors and space are critical for minimising potential negative effects. For example, integration potentially ensures that disaster risk management at the local level is aligned with national and regional mechanisms (Allan 2017), but also across civil emergencies, infrastructure maintenance and human rights (APFM 2006). Such coordination is challenging but essential if trade-offs are to reflect the relative values of each element.

Physical and administrative boundaries rarely coincide for deltas. Thus, decisions made outside, or in different areas of, the delta may have disproportionate effects (see, for example, the impact of dams on sediment transport in Chapter 6). Policymaking and the relevant legal architecture in relation to the management of water resources and floods, for example the European Water Framework Directive (Article 3, European Parliament and Council 2000), generally express balance between upstream and downstream rights and interests. More specifically, delta nations such as the Netherlands, Vietnam and Bangladesh, have international agreements in place that make their planning processes more robust and flexible (Seijger et al. 2017). However, in most cases, this does not happen (for example, India), as there is no agreement at basin level on how the river is to be managed (Global Water Partnership 2000). This has the effect of limiting or skewing the trade-offs and reducing options that can be considered. Lack of data-sharing mechanisms and limitations on basin management, whether at national or international scales, may result in a similar situation (Gerlak et al. 2011).

Trade-offs that maximise benefits and minimise disadvantages would appear to be desirable. But governments typically do not have complete autonomy in decision-making, and are constrained by international obligations. For example, the rights of co-riparian states under international law relating to transboundary waters (United Nations 1997), the rights of women (United Nations 1966), the views of relevant stakeholders (UNECE 1998), protection of the environment (e.g. Ramsar 1971) and human rights more generally. Such international frameworks also inform those making policy regarding deltas about best practice. The Sendai Framework on Disaster Risk Reduction (United Nations 2015) for example, illuminates how trade-offs can be put into effect. Human rights considerations have broader impacts for policymaking and the assessment of trade-offs as they may oblige policymakers to make public the considerations that have not only been considered in policy development and implementation, but more importantly to define those people who have rights to be involved in the making of those policies. For deltas in particular, consideration of the aims, obligations and challenges of global issues, such as climate change, can also affect decision-making (Hoegh-Guldberg et al. 2018).

In the end, governments that are seeking to develop policy for deltas and their inhabitants will be framed, informed and unavoidably influenced by factors over which they may have little control. However, delta planning processes, principally affecting flood and water management infrastructure, remain likely to be more effective if they are nested within flood and water resource management governance which extends across whole basins.

10.3 Assessing Trends and Trade-Offs Under Plausible Delta Futures

Trends and trade-offs are, it is suggested, symptomatic of delta development during the Anthropocene. Hence their identification and evaluation play an important role in understanding the future of deltas thus enabling evidence-based decision-making and planning (Anderies et al. 2007; Daw et al. 2016). Integrated Assessment Modelling captures the main elements and characteristics of the system in a framework and allows the testing of a range of scenarios. Trade-offs are identified by comparing scenarios and identifying the winners and losers of the changes (Daw et al. 2011; Chapman and Darby 2016). Thus, integrated modelling can help to identify plausible trends, and locate, estimate and highlight the risks or benefits (including previously unforeseen) associated with proposed or unintended changes for future delta development.

Here a novel integrated assessment model, the Delta Dynamic Integrated Emulator model (ΔDIEM) is used. Originally developed for coastal Bangladesh in the GBM Delta (Nicholls et al. 2018), it allows the exploration of the interactive relationships between natural and socio-economic processes under a range of scenarios and policy options to identify both trends and trade-offs (Nicholls et al. 2016; Hutton et al. 2018; Lázár et al. 2018b). The following sections illustrate trade-offs using different approaches and scenarios in each case; please note that comparison between examples is therefore not possible.

Within ΔDIEM, contrasting scenarios can be constructed from the model variables to explore the effects on the well-being of the local population and productivity. Table 10.1 shows the main components

Table 10.1 Summary of assumed changes from present (2015) to 2100 for coastal Bangladesh within the two selected scenarios

	Scenario descriptions	
	Positive world	Negative world
Sea-level rise	Slow—54 cm (26 cm in 2050)	Fast—148 cm (61 cm in 2050)
Air temperature	+3.7 °C	+4.1 °C
Other climate	Moderately drier climate (−15% annual precipitation compared to mean of 2000–2015; −17% total monsoon rain and −3% total dry season rainfall)	Wetter climate (+10% annual precipitation compared to mean of 2000–2015) with more extremes: wetter monsoon (+2%) and drier dry seasons (−18%)
Population	Maintained at 14 million	Decreasing to seven million, reflecting net outmigration
Embankment maintenance	Good maintenance—maintained embankments	Poor maintenance—slow deterioration in embankment height (−3 cm per year)
Farming and agriculture	New crop types increase production (+20%)	Traditional crop types, no growth in production
Manufacturing and services	Fast growth (212% by 2100 or 2.5% per year)	Slow growth (53% by 2100 or 0.62% per year)

of two extreme scenarios used to examine potential trends in coastal Bangladesh within the GBM Delta.

Figure 10.1 compares the expected trends for six indicators over this century associated with the selected scenarios. It is clear that each scenario has both positive and negative effects. For example, using the end numbers of the scenario configurations identified above, embankment deterioration, higher sea-level rise and more seasonality in river flows, the total inundated area increases by approximately 70% under the Negative scenario. In the alternative future with fewer extremes and better water resources management, the total inundation area does not change significantly.

There are environmental trade-offs in both worlds. In the Positive scenario, less flooding increases the soil salinity by about 30%, due to

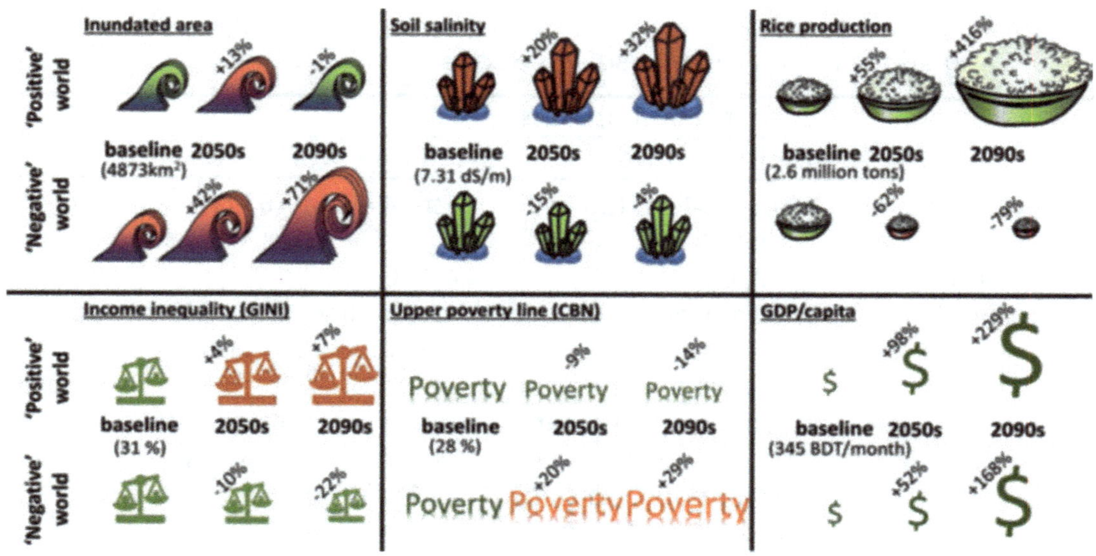

Fig. 10.1 Contrasting trends for coastal Bangladesh between the 'negative' and 'positive' scenarios (baseline: mean of 2005–2015)

less flushing effects, but this is compensated with salt tolerant crops and better management techniques, resulting in a fourfold net increase in rice productivity. Thus, food security and economy are not impacted, but environmental quality is degraded. In the opposite, Negative world, soil salinity slightly decreases due to the annual monsoon flushing of the salt out of soil. However, rice production collapses by about 80% due to the frequent flood damage, elevated temperature limitations and lower potential yields of traditional rice varieties compared to higher yielding varieties of the Positive world. Also, while environmental quality improves with more sediment deposition and lower salinity levels under the Negative world, traditional farming practices could have negative implications on the local economy. Of course, there are plausible futures which sit between these scenarios and these can be explored by the use of the modelling.

There are similar contradictions within and between income and inequality trends. Most noteworthy is that while GDP per capita increases significantly in the Positive world so does inequality indicating that the income gap within society widens. This increase in inequality lies in unequal land sizes and off-farm capital investments. The more land one owns, the more profit it makes under good economic conditions, whereas the economic growth of the poor and landless is slower. Hence the income gap widens. Income inequality is significantly lower in the Negative world where a lower per capita growth is experienced, but it is more equitably distributed. When the simulated household consumption rates are compared with the extrapolated trend of the upper poverty line of Bangladesh (BBS 2011), an increasing percentage of the Negative world population remains classified as in poverty. This indicates that poverty can persist even when the GDP per capita growth occurs, particularly where manufacturing and services form an important aspect of the local economy. Thus, targeted policies and development programs are required to provide a safety net and support for the poorest households and regions in any of the futures.

10.4 Policy Driven Trade-Offs

Variations in potential futures can also be strongly related to governance decisions and policy focus. Policy is generally set in relation to a single issue with little capacity for consideration of potential side effects or trade-offs with other sectors of society. This section presents a comparison of the effects of three contrasting policy examples (Table 10.2) with an Economic Development, a Social Welfare and an Environmental Sustainability focus against a baseline scenario describing a continuation of current trends. The **Economic Development policy** focuses on promoting productivity. Agriculture is promoted with progressively improving crop varieties. Irrigation use, irrigation efficiency and fertiliser use are also maximised. Fishing intensity is increased to get more economic benefits. Embankments are maintained to reduce the possibility of flooding. Population numbers are the same as the baseline scenario. For the **Social Welfare policy**, selected options aim to support households residing in the delta. The population remains at 14 million assuming that as a result of the policies net outmigration is zero. Agriculture retains the use of existing irrigation and fertiliser practices, and utilises new crops following research and development activities after 2050 to maximise income. In addition, agriculture costs are subsidised (25%) and household expenses are reduced by 25%

Table 10.2 Scenario summary of the policy trade-off simulations (by 2100). Cells with shading highlights the deviation from the Baseline scenario assumptions

	Baseline (current trends)	Economic development	Social welfare	Environmental sustainability
Climate	A warmer (+3.7 °C) and moderately drier (–13% precipitation) climate			
Sea-level rise	Fast—148 cm by 2100			
Polders	Maintained existing			Deteriorate (–3 cm/yr)
Tidal river management	No			Yes (2020 for 5 years; all polders)
Population	Decreasing to 10 million	Decreasing to 10 million	Maintained at 14 million	Decreasing to 7 million
Land cover	Present land cover, slightly increasing agriculture at the expense of mangroves	Better zoning in land use, slight decrease in agriculture, but increasing urbanisation	Present land cover, slightly increasing agriculture at the expense of mangroves	Better zoning in land use, slight decrease in agriculture, but increasing urbanisation
Economy	Moderate growth (off-farm: 106%, ES-based: 53%)	Fast growth (off-farm: 212%, ES-based: 106%)	Moderate growth (off-farm: 106%, ES-based: 53%) BUT lower household costs (–25%)	Moderate growth (off-farm: 106%, ES-based: 53%)
Subsidies	No	No	Agriculture costs (-25%)	No
Agriculture crops	Present day crops	new crop types increased production (+20%)	Present day crops	Traditional crop types, no growth in production
Irrigation	Present practices	everywhere	Present practices	No irrigation
Irrigation efficiency	90%	100%	90%	n/a
Fertiliser use	Present practices	Present practices	Present practices	No fertiliser
Fishing intensity	Twice the sustainable level	Four times the sustainable level	Sustainable level	Sustainable level

throughout the simulations to relieve the households financially and make life easier, so that they do not migrate away. Fishing intensity is reduced to provide a long-term sustainable fishing income. Under the **Environmental Sustainability policy**, the aim is to restore as far as possible the natural condition of the delta. The dyke system is allowed to slowly deteriorate over this century increasingly allowing flooding, and thus increasing sedimentation. In addition, Tidal River Management for all polders during the 2020s enable sediment accumulation: further interventions are not needed as flooding and thus sedimentation will intensify over time as embankments deteriorate. Farming practices go back to traditional crops and methods and do not use irrigation or fertiliser. Fishing intensity is reduced to a sustainable level. Population numbers fall to seven million by the end of the century. The specific options selected within ΔDIEM for each policy are summarised in Table 10.2; climate and sea-level scenarios remain constant.

The potential trade-offs are shown as the percentage of difference for six selected indicators when compared to the Baseline scenario. The results are surprising (Fig. 10.2). By 2050, the changes compared to Baseline are relatively small. Soil salinity is drastically reduced under the Environmental Sustainability scenario as a result of the deteriorating embankments and thus more frequent flooding and monsoon flushing of salt from the soil. However, soil salinity trends will not change as a result of the other scenarios. Rice productivity is maintained when the irrigation is maximised, but in all other cases, the productivity will fall. Returning to traditional farming practices would leave the region with about 70% less rice. Income inequality (GINI index) is not impacted

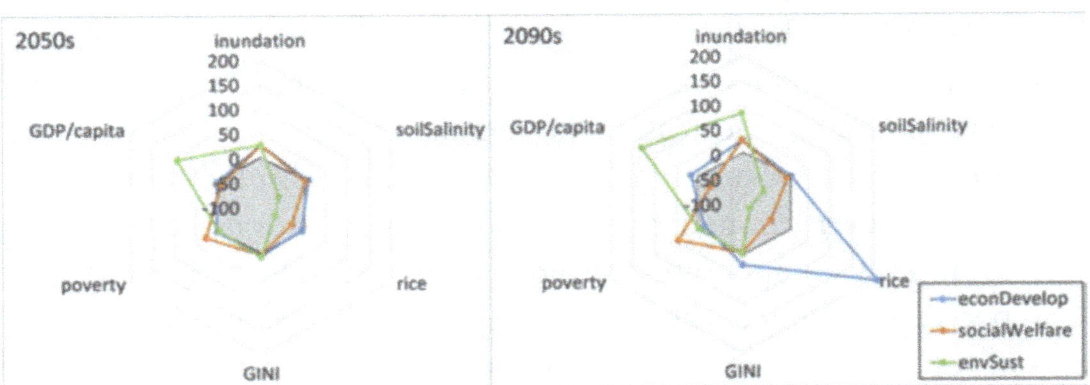

Fig. 10.2 Trade-offs resulting from different policy focuses in coastal Bangladesh. The values show the relative change (percent) compared to the baseline simulation (i.e. current trends). The grey shaded area indicates negative change

significantly by these governance changes. Poverty however, would increase under the Social Welfare scenario (+25%). Finally, GDP per capita would sharply increase under the Environmental Sustainability scenario (+90%).

By the end of the century, these trends are amplified. Flooding intensifies under the Environmental Sustainability scenario due to the disappearance of the dyke system in coastal Bangladesh. The soil salinity does not deviate further from the Baseline. However, rice productivity multiplies under the Economic Development scenario (+200%) and falls to an even lower level under the Environmental Sustainability scenario (−80%). Inequality increases under the Economic Development scenario (+20%), but poverty is reduced under the same scenario by ten percent. However, poverty increases under the Social Welfare scenario by approximately 50% (compared to Baseline). As a result of this poverty increase, GDP/capita falls under the Social Welfare scenario, but increases under Economic Development (~20%) and Environmental Sustainability (130%).

The results described above provide somewhat counter-intuitive results to conventional wisdom concerning the outcome of policy strategies. Why would the Social Welfare scenario result in higher poverty rates and why would the Environmental Sustainability scenario generate higher GDP per capita results than the Economic Development scenario? Based on the model results, a key factor lies in the assumptions about population. The Environmental Sustainability scenario assumes a rapidly declining population and a land redistribution. This means that as the population size decreases, the remaining farmers would accumulate more productive land and farm size increases through land consolidation. Thus, even though the rice productivity is drastically reduced, larger farm sizes compensate for the loss of yield resulting in a higher GDP per capita value. The scenario also has better biodiversity as a result of the reconnected land to river, lower soil salinity and sustainable fishing habits. However, more frequent and extensive flooding can be expected.

The Economic Development scenario solely focuses on economic gains, thus exploiting the environment with higher fishing efforts and more intrusive farming practices. Rice productivity increases significantly, but population size remains the same as in the Baseline. This results in only a moderate increase in GDP per capita, as the share from this significant increase in productivity mainly benefits the large landowners. As a result, poverty slightly decreases, but inequality noticeably increases by the end of the century.

At the heart of the Social Welfare scenario is the aim to support rural livelihoods and welfare through subsidies on household expenses and agriculture costs. However, this aim is not achieved in the simulation. The population size remains at today's level, but this results in smaller farm sizes. Rice productivity declines slightly due to slightly increased flooding and the new subsidies are not sufficient to compensate for the loss in income and small farm size. Since more landless and small landowners remain, the overall poverty rate increases and the GDP per capita falls. Thus, investment in social capital and land consolidation seem to be key to make the coastal zone socio-economically more sustainable. Of course, in undesirable circumstances, there is likely to be an increase in migratory behaviour and possibly a greater uptake of workers into the growing service sector, a subject of ongoing research. In conclusion, there is a delicate balance between livelihood potential, population size, welfare programs and environmental hazard mitigation that national planning has to carefully consider and which can be supported by integrative model simulations.

The results highlight the interlinked natural-socio-economic delta processes in action. Even though the results are illustrative, they are also robust. However, these still need to be used with care due to assumptions in the model setup. For example, there is an assumed land consolidation that might not happen in coastal Bangladesh, because land is an important safety net for agriculturally-dependent populations (Toufique and Turton 2002). Economic trends are also assumed and future changes are very uncertain especially under climate change and when flooding patterns significantly change (see Chapter 8). Thus, to inform policy, improved understanding of the basis for assumptions is needed, with a more robust scenario testing exercise, particularly as they may vary spatially and temporally. Changes outside coastal Bangladesh may also need to be considered.

10.5 Spatial Trade-Offs

Trade-offs can also occur spatially. An example is shown in Fig. 10.3, which illustrates the flooding and socio-economic consequences of new polder development in coastal Bangladesh (Chapman et al. 2019). Under the current situation, the North-East part of the study area is regularly flooded in the monsoon season (Fig. 10.3a). Hence, the construction of new embankment and polders is proposed to reduce flooding and better manage water levels. However, hydrodynamic modelling shows that when these new polders are implemented, the neighbouring Western areas experience more extensive and deeper flooding (Fig. 10.3b). This is reflected in the decrease in agricultural output and increase in households in poverty for this area. However, these losses are significantly less than the gains experienced in the newly protected area, so there is a net benefit to Bangladesh. Although this does not include construction costs, it is clear that the economic benefits outweigh the damages; yet, significant livelihood and well-being changes can be expected in the neighbouring areas. The identification of this spatial trade-off allows consequences to be pro-actively managed, possibly with additional financial support to compensate for losses, new flood defence or training for new livelihoods.

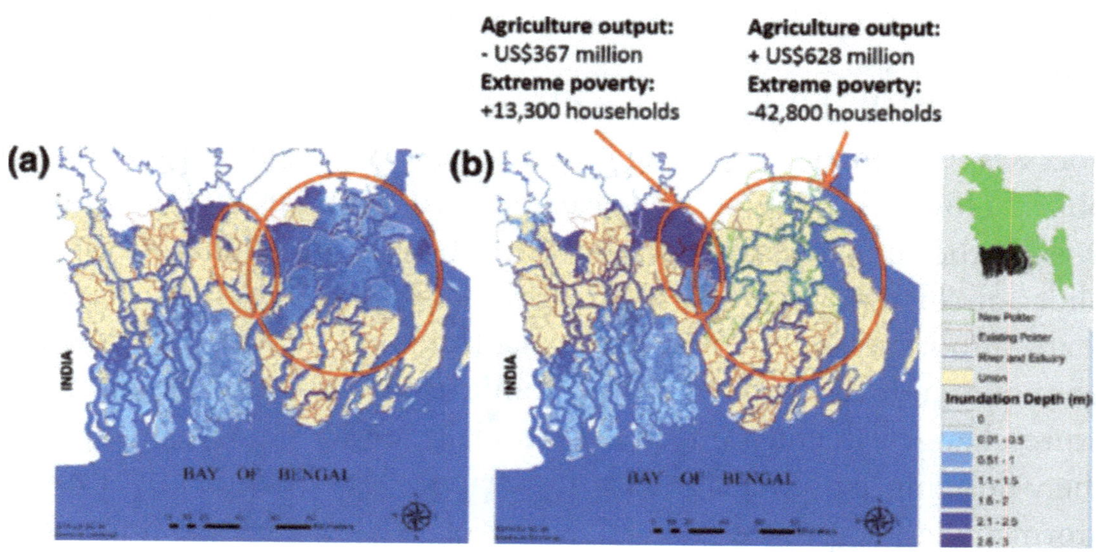

Fig. 10.3 Risk transfer due to new polders in coastal Bangladesh **a** existing polders and **b** existing with new South-Central polders (Adapted from Haque and Nicholls [2018] under CC BY 4.0)

However, there are some other longer-term issues that need to be considered. For example, additional polders will increase the sediment deficit on the land (Chapter 5) resulting in accelerated subsidence and waterlogging. Also, the reduced flood risk will result in an increase in land value, and the poorest inhabitants might be forced off their land reducing their chance for an increased well-being. The social structure of these areas could therefore change due to these improvements and result in within-community conflicts. In addition, the poor who sell their land for short-term capital gain need to move elsewhere, potentially increasing the poverty rate and local issues (e.g. infrastructure overload) of the receiving area. In Bangladesh, the capital Dhaka is the main destination of such migration (Chapter 7).

10.6 Household Adaptation Response to Change

For policy development, in addition to identification of trade-offs such as those in Sect. 10.3, it is also beneficial to consider how the local population might respond to such changes at household level. As discussed in Chapters 7 and 9, people, and the households in which they reside, constantly adapt to changing natural and socio-economic conditions. This is a feature of life everywhere, but the high natural and social pressures in populous deltas accentuate these processes.

Based on analysis of household surveys in three deltas (see Chapters 2–4), an Integrated Bayesian Adaptation model is able to quantify the influence of different policy directions on household decision-making (Lázár et al. 2015). This model considers the characteristics of the household and the environment to quantify the possibility of the household to adapt its present behaviour.

Five broad categories of household adaptation options are simulated (Lázár et al. 2018a): 1. A **financial** change alters or supplements the household finances by getting a loan, insuring the livelihood activities and assets, applying for an NGO or government support or joining a cooperative; 2. A **structural** change can improve the house or make it climate resilient, trees might be planted around the house to provide shade, firewood or additional protection against wind damage, or the household might start using community shelters when disasters are imminent; 3. The household can **modify or intensify** the existing livelihood by start using fertiliser, irrigation, new productive assets and more tolerant crops, but it can also hire additional labour or participate in training courses; 4. **Diversifying** the existing livelihood might include growing a wider range of crops in multiple seasons, engaging for example in mixed farming/fishing activities, or enabling women to take up work outside the house; and finally, 5. the household might decide to send a household member away to alternate labour markets or for non-economic reasons such as family issues or education (i.e. **migrate**).

In this example, this Integrated Bayesian Adaptation model is used to simulate the effect of six plausible policy directions on household decisions. As discussed in (Suckall et al. 2018) and Sect. 9.5 of this book, these six plausible adaptation policy directions (APDs) were conceptualised representing contrasting governance contexts: A. **Minimum Intervention** (low investment/low commitment to policy change): no-regrets strategy where the lowest cost adaptation policies are pursued to protect citizens from some climate impacts; B. **Capacity Expansion** (high investment/low commitment to policy change) encourages climate-proof economic growth, but does not seek to make significant change to the current structure of the economy; C. **Efficiency Enhancement** (medium investment/medium commitment to policy change) is an ambitious strategy that promotes adaptation consistent with the most efficient management and exploitation of the current system, and best utilising ecosystem services to enhance livelihoods and well-being under climate change; D. **System Restructuring** (high investment/high commitment to change) embraces pre-emptive fundamental change at every level in order to completely transform the current social and ecological system: D1. Protect, broadly following the Dutch model with a high level of engineered protection, D2. Accommodate, live and work with nature principle, D3. Retreat population and infrastructural relocation.

Each of the above investment policy directions was broken down to a set of assumptions describing the changes of household and environmental characteristics as a result of the presence of APDs. During the model simulation, household and environmental characteristics are altered and the new adaptation choices are recorded. When comparing the baseline and new adaptation choices, the percent changes in household decisions can be quantified and compared (Fig. 10.4, N.B. the direction of the change can be positive or negative, and this is not shown). The most common adaptation responses for both male- and female-headed households are financial, where they seek alternative income sources, and structural, which is primarily emergency response to an environmental hazard. These broad trajectories indicate an improving household welfare.

Migration has limited appeal as an adaptation option with only female-headed households living in the Bangladeshi GBM and male-headed in the Indian administered part of the GBM Delta (also termed the Indian Bengal Delta [IBD]), considering this as a viable option perhaps due to the higher level of protection and other social benefits. This indicates that, where possible, the general preference is for the population to remain in situ and adapt. The frequency of livelihood adaptations is also unlikely to increase in the future under these APDs. Overall, the results indicate that better policies can positively influence

Male headed	GBM		Mahanadi	Ghana
	Bangladesh	IBD		
Financial	***	***	***	*
Structural	***	***	***	**
Intensify	**	*	**	**
Diversify	**	*	*	*
Migrate	*	*	*	*

Female headed	GBM		Mahanadi	Ghana
	Bangladesh	IBD		
Financial	***	***	**	*
Structural	***	***	***	*
Intensify	*	*	*	*
Diversify	*	*	*	*
Migrate	**	*	*	*

Legend:

no (<=5% change)
moderate (5-20% change)
high (>20% change)

*** (>66% prevalence)
** (34-66% prevalence)
* (<34% prevalence)

Fig. 10.4 Prevalence and the level of influence of adaptation policy directions on gendered household decisions in specified deltas in situ livelihoods and well-being, but migration trends are unlikely to be changed until there is a significant shift in well-being and as a result, the mindset of households changes significantly.

10.7 Conclusion

Natural deltas are complex systems with many interacting factors and that operate on diverse timescales, from seasons through to geological time. However, in the Anthropocene, human actions have become dominant and the complexity of delta system processes has grown significantly. As a consequence, the assessment of trade-offs and the balance between those that gain or lose are becoming highly important for policy development. Trade-offs between and within the socio-economic and natural systems are inevitable due to the competing interests and limited understanding of causal effects.

Deltas are key social and economic hubs and they are critical to the livelihoods and well-being of their resident and neighbouring populations. Deltas are also areas where any development is exposed to significant risks (Chapter 8) with environmental hazards growing due to climate change and other factors: submergence by sea-level rise is an obvious and widely discussed threat (see Chapter 6 and Wong et al. 2014, for example). Historical analogues might not be valid to infer the future due to feedbacks and system shifts. There is, therefore, a pressing need to understand the behaviour and co-evolution of the coupled human-natural systems and to simultaneously develop and manage these systems in ways that are positive for society while sustaining the environment. Integrated models can explore these interactions through policy and management options, identifying trade-offs and their possible consequences, an important contribution for key global policy goals such as the internationally-agreed Sustainable Development Goals (Hutton et al. 2018). They are also consistent with moves to adaptive delta planning such as the Bangladesh Delta Plan 2100.

The examples discussed in this chapter demonstrate how climate and environmental change is both an important driver and limiting factor for delta development. However, technological development and good governance can result in positive delta management. What kind of development is a choice for governments and associated decision-makers. Economic growth is often the preferred objective, but this can result in higher segregation within communities, large scale environmental degradation and other ill-considered consequences. Effective balancing of trade-offs at basin level is enhanced by the existence of water resource management and flood management frameworks at this scale, and along with efficient data sharing arrangements, potentially lead to more enlightened delta management.

Actions of resident populations are crucial to the sustainability of future deltas; their relationship with the delta system, their use of resources and development of infrastructure inevitably influence environmental and livelihood potential changes and adaptation options. Thus, actions by individuals and households can either support or negate larger scale policy decisions. Combining the top-down policy influences with the bottom-up household actions is the way forward in integrated modelling of delta systems.

In many deltas, major investment in coastal protection provides only a short term fix to rising sea levels. Ultimately, all deltas will need to face the uncomfortable truth that any further large-scale infrastructure development will be highly costly, leading to potential lock-in and requiring continuous and substantial capital and maintenance funds. Thus, when planning for the long-term beyond the next 50 years, continuous adaptation at both national level (Haasnoot et al. 2013) and at local-scale in livelihood adaptation, investment in human capital, is more likely to secure long-term delta sustainability. Incorporating natural processes, such as space for river and tidal river management, holds high promises and should be prioritised wherever is possible in delta planning and management.

It is clear that managing coupled human and natural systems inevitably involves trade-offs, with winners and losers. There is no win-win situation for both delta systems and the use of its ecosystem services. Delta planning, thus, remains an intractable challenge with no easy solution. Looking to the future, these types of integrated analysis tools (e.g. ΔDIEM) provide new perspectives, hold great promise and also serve a fundamental and growing need.

References

Allan, A. (2017). Legal aspects of flood management. In A. Rieu-Clarke, A. Allan, & S. Hendry (Eds.), *Routledge handbook of water law and policy*. London, UK: Earthscan.

Amoako-Johnson, F., Hutton, C. W., Hornby, D., Lázár, A. N., & Mukhopadhyay, A. (2016). Is shrimp farming a successful adaptation to salinity intrusion? A geospatial associative analysis of poverty in the populous Ganges–Brahmaputra–Meghna Delta of Bangladesh. *Sustainability Science, 11*(3), 423–439. https://doi.org/10.1007/s11625-016-0356-6.

Anderies, J. M., Rodriguez, A. A., Janssen, M. A., & Cifdaloz, O. (2007). Panaceas, uncertainty, and the robust control framework in sustainability science. *Proceedings of the National Academy of Sciences, 104*(39), 15194–15199. https://doi.org/10.1073/pnas.0702655104.

APFM. (2006). *Legal and institutional aspects of integrated flood management* (APFM Technical Document No. 2). Associated Program on Flood Management, World Meteorological Organization, Geneva, Switzerland.

BBS. (2011). *Report of the household income and expenditure survey 2010*. Bangladesh Bureau of Statistics (BBS), Ministry of Planning, Government of the People's Republic of Bangladesh, Dhaka, Bangladesh. http://203.112.218.65:8008/WebTestApplication/userfiles/Image/LatestReports/HIES-10.pdf. Last accessed 12 November 2018.

Chapman, A., & Darby, S. (2016). Evaluating sustainable adaptation strategies for vulnerable mega-deltas using system dynamics modelling: Rice agriculture in the Mekong Delta's An Giang Province, Vietnam. *Science of the Total Environment, 559*, 326–338. https://doi.org/10.1016/j.scitotenv.2016.02.162.

Chapman, A., Noman, M. Z. N., Lázár, A. N., Nicholls, J. R., Haque, A., Hussain, M. M., et al. (2019). Supporting adaptation and development policy making in deltas: Application of an integrated assessment model in coastal Bangladesh. *Journal of Environmental Planning and Management* (in review).

Daw, T., Brown, K., Rosendo, S., & Pomeroy, R. (2011). Applying the ecosystem services concept to poverty alleviation: The need to disaggregate human well-being. *Environmental Conservation, 38*(4), 370–379. https://doi.org/10.1017/S0376892911000506.

Daw, T. M., Hicks, C. C., Brown, K., Chaigneau, T., Januchowski-Hartley, F. A., Cheung, W. W. L., et al. (2016). Elasticity in ecosystem services:

Exploring the variable relationship between ecosystems and human well-being. *Ecology and Society, 21*, 2. https://doi.org/10.5751/ES-08173-210211.

Day, J. W., Agboola, J., Chen, Z., D'Elia, C., Forbes, D. L., Giosan, L., et al. (2016). Approaches to defining deltaic sustainability in the 21st century. *Estuarine, Coastal and Shelf Science, 183*, 275–291. https://doi.org/10.1016/j.ecss.2016.06.018.

European Parliament and Council. (2000). Water Framework Directive (2000/60/EC). *Official Journal of the European Communities*.

Gerlak, A. K., Lautze, J., & Giordano, M. (2011). Water resources data and information exchange in transboundary water treaties. *International Environmental Agreements: Politics, Law and Economics, 11*(2), 179–199. https://doi.org/10.1007/s10784-010-9144-4.

Global Water Partnership. (2000). *Integrated water resources management* (TAC Background Paper 4). Stockholm, Sweden: Global Water Partnership. https://www.gwp.org/globalassets/global/toolbox/publications/background-papers/04-integrated-water-resources-management-2000-english.pdf. Last accessed 12 November 2018.

Haasnoot, M., Kwakkel, J. H., Walker, W. E., & ter Maat, J. (2013). Dynamic adaptive policy pathways: A method for crafting robust decisions for a deeply uncertain world. *Global Environmental Change, 23*(2), 485–498. https://doi.org/10.1016/j.gloenvcha.2012.12.006.

Haque, A., & Nicholls, R. J. (2018), Floods and the Ganges-Brahmaputra-Meghna Delta. In R. J. Nicholls, C. W. Hutton, W. N. Adger, S. E. Hanson, M. M. Rahman, & M. Salehin (Eds.), *Ecosystem services for well-being in deltas: Integrated assessment for policy analysis* (pp. 147–159). Cham: Springer. http://dx.doi.org/10.1007/978-3-319-71093-8_8.

Hoegh-Guldberg, O., Jacob, D., & Taylor, M. (2018). Impacts of 1.5°C global warming on natural and human systems. In *Global warming of 1.5 °C—An IPCC special report on the impacts of global warming of 1.5 °C above pre-industrial levels and related global greenhouse gas emission pathways, in the context of strengthening the global response to the threat of climate change, sustainable development, and efforts to eradicate poverty* (246 pp.). Incheon, Republic of Korea: Intergovernmental Panel on Climate Change (IPCC). http://report.ipcc.ch/sr15/pdf/sr15_chapter3.pdf. Last accessed 13 November 2018.

Hutton, C. W., Nicholls, R. J., Lázár, A. N., Chapman, A., Schaafsma, M., & Salehin, M. (2018). Potential trade-offs between the Sustainable Development Goals in coastal Bangladesh. *Sustainability, 10*(4), 1008. http://dx.doi.org/10.3390/su10041108.

Lázár, A. N., Nicholls, R. J., Payo, A., Adams, H., Mortreux, C., Suckall, N., et al. (2015). *A method to assess migration and adaptation in deltas: A preliminary fast-track assessment* (Deltas, Vulnerability and Climate Change: Migration and Adaptation [DECCMA] Working Paper). Southampton, UK: DECCMA Consortium. https://generic.wordpress.soton.ac.uk/deccma/resources/working-papers/. Last accessed 3 September 2018.

Lázár, A. N., Adams, H., de Campos, R. S., Vincent, K., Nicholls, J. R., & Adger, W. N. (2018a). *How gender affects household decisions to adapt or migrate due to environmental risk: A Bayesian network model for coastal Bangladesh* (DECCMA Working Paper). Deltas, Vulnerability and Climate Change: Migration and Adaptation, IDRC Project Number 107642.

Lázár, A. N., Payo, A., Adams, H., Ahmed, A., Allan, A., Akanda, et al. (2018b). Integrative analysis applying the delta dynamic integrated emulator model in south-west coastal Bangladesh. In R. J. Nicholls, C. W. Hutton, W. N. Adger, S. Hanson, M. M. Rahman, & M. Salehin (Eds.), *Ecosystem services for well-being in deltas: Integrated assessment for policy analysis*. Cham, Switzerland: Springer. http://dx.doi.org/10.1007/978-3-319-71093-8_28.

Nicholls, R. J., Hutton, C., Adger, W. N., Hanson, S. E., Rahman, M. M., & Salehin, M. (Eds.). (2018). *Ecosystem services for well-being in deltas: Integrated assessment for policy analysis*. London, UK: Palgrave Macmillan.

Nicholls, R. J., Hutton, C. W., Lázár, A. N., Allan, A., Adger, W. N., Adams, H., et al. (2016). Integrated assessment of social and environmental sustainability dynamics in the Ganges-Brahmaputra-Meghna Delta, Bangladesh. *Estuarine and Coastal Shelf Science, 183*, 370–381. https://doi.org/10.1016/j.ecss.2016.08.017.

Overeem, I., & Syvitski, J. P. M. (2009). *Dynamics and vulnerability of delta systems* (LOICZ Reports and Studies 35). Geesthacht, Germany: GKSS Research Center.

Paavola, J., Gouldson, A., & Kulvankova-Oravska, T. (2009). Interplay of actors, scales, frameworks and regimes in the governance of biodiversity. *Environmental Policy and Governance, 19*(3), 148–158. http://dx.doi.org/10.1002/eet.505.

Ramsar. (1971, January 30–February 3). *The final act of the international conference on conservation of wetlands and waterfowl*. Ramsar, Iran.

Seijger, C., Douven, W., van Halsema, G., Hermans, L., Evers, J., Phi, H. L., et al. (2017). An analytical framework for strategic delta planning: Negotiating consent for long-term sustainable delta development. *Journal of*

Environmental Planning and Management, 60(8), 1485–1509. https://doi.org/10.1080/09640568.2016.1231667.

Steffan-Dewenter, I., Kessler, M., Barkmann, J., Bos, M. M., Buchori, D., Erasmi, S., et al. (2007). Tradeoffs between income, biodiversity, and ecosystem functioning during tropical rainforest conversion and agroforestry intensification. *Proceedings of the National Academy of Sciences, 104*(12), 4973–4978. https://doi.org/10.1073/pnas.0608409104.

Suckall, N., Tompkins, E. L., Nicholls, R. J., Kebede, A. S., Lázár, A. N., Hutton, C., et al. (2018). A framework for identifying and selecting long term adaptation policy directions for deltas. *Science of the Total Environment, 633,* 946–957. https://doi.org/10.1016/j.scitotenv.2018.03.234.

Toufique, K. A., & Turton, C. (2002). *Hands not land: How livelihoods are changing in rural Bangladesh.* Dhaka, Bangladesh: Bangladesh Institute of Development Studies.

UNECE. (1998). *Convention on access to information, public participation in decision-making and access to justice in environmental matters (Aarhus Convention).* Aarhus, Denmark, 25 June 1998, entered into force 2001. http://www.unece.org/fileadmin/DAM/env/pp/documents/cep43e.pdf.

United Nations. (1966). *International covenant on economic social and cultural rights.*

United Nations. (1997). *Convention on the law of the non-navigational uses of international watercourses* on the 21 May 1997, entered into force 17 August 2014.

United Nations. (2015). *Sendai framework for disaster risk reduction 2015–2030.*

Wolman, M. G., & Gerson, R. (1978). Relative scales of time and effectiveness of climate in watershed geomorphology. *Earth Surface Processes, 3*(2), 189–208. https://doi.org/10.1002/esp.3290030207.

Wong, P. P., Losada, I. J., Gattuso, J.-P., Hinkel, J., Khattabi, A., McInnes, K. L., et al. (2014). Coastal systems and low-lying areas. In C. B. Field, V. R. Barros, D. J. Dokken, K. J. Mach, M. D. Mastrandrea, T. E. Bilir, et al. (Eds.), *Climate change 2014: Impacts, adaptation, and vulnerability. Part A: Global and sectoral aspects* (pp. 361–409). Contribution of Working Group II to the Fifth Assessment Report of the Intergovernmental Panel on Climate Change. Cambridge, UK and New York, NY: Cambridge University Press.

11

Sustainable Deltas in the Anthropocene

Robert J. Nicholls, W. Neil Adger, Craig W. Hutton, Susan E. Hanson, Attila N. Lázár, Katharine Vincent, Andrew Allan, Emma L. Tompkins, Iñaki Arto, Md. Munsur Rahman, Sugata Hazra and Samuel Nii Ardey Codjoe

11.1 Introduction

This book examines the recent development of selected populous deltas from a variety of perspectives—sediment budgets, vulnerability hotspots, settlement and migration, the delta economy, household adaptation and delta-level adaptation. Building on these analyses, this chapter extracts key lessons for delta development and management through the early twenty-first century and beyond. Some of the emerging trends seem almost inevitable, such as declining sediment supplies, while others will

R. J. Nicholls (✉) · S. E. Hanson · A. N. Lázár
School of Engineering, University of Southampton, Southampton, UK
e-mail: r.j.nicholls@soton.ac.uk

W. N. Adger
Geography, College of Life and Environmental Sciences,
University of Exeter, Exeter, UK

R. J. Nicholls et al. (eds.), *Deltas in the Anthropocene*,
https://doi.org/10.1007/978-3-030-23517-8_11

depend strongly on the choices that are made, such as how delta populations are enabled to adapt. Hence, what are the possible trajectories for delta development and how can the more desirable ones be achieved?

Deltas are home to 1 in 14 of the global population: hence a key issue is the implications of delta science for the sustainability and persistence of deltas as geomorphic, ecological, economic and social systems. Many chapters in this book highlight how deltas are dynamic in all these aspects today and change seems inevitable. As deltas are fundamentally geomorphic features, their geomorphic persistence is a core prerequisite for delta sustainability in the long-term (Syvitski et al. 2009; Giosan et al. 2014; Day et al. 2016). Rapid socio-economic changes are also a feature of

---

C. W. Hutton
GeoData Institute, Geography and Environmental Science, University of Southampton, Southampton, UK

K. Vincent
Kulima Integrated Development Solutions, Pietermaritzburg, South Africa

A. Allan
School of Law, University of Dundee, Dundee, UK

E. L. Tompkins
Geography and Environmental Science, University of Southampton, Southampton, UK

I. Arto
bc³—Basque Centre for Climate Change, Bilbao, Spain

M. M. Rahman
Institute of Water and Flood Management, Bangladesh University of Engineering and Technology, Dhaka, Bangladesh

S. Hazra
School of Oceanographic Studies, Jadavpur University, Kolkata, India

S. N. A. Codjoe
Regional Institute for Population Studies, University of Ghana, Legon-Accra, Ghana

Anthropocene deltas. What are the implications of these different types of change and their different timescales? Economic development gives the opportunity for deltas to prosper and increase the welfare of their whole population over the coming few decades. Given current low levels of income of populations in many agriculture-dominated delta regions, such development is essential. However, in the decade to century time-scale, how will the bio-geomorphic constraints operate and can populations adapt to them? The evidence in this book shows that the twenty-first century will continue to transform landscapes as deltas become more engineered, urbanised and more central to economic activity in their wider regions. Yet the decisions taken in the next decade will set in train pathways that are potentially irreversible. The benefits are increasingly clear of conserving natural capital and conserving ecosystem services in the present day to keep options open for nature-based adaptation into a sustainable and good Anthropocene for deltas (Bennett et al. 2016).

The chapter is structured around the three specific questions asked above. First, the Anthropocene transition in deltas is synthesised. This, second, forms the basis of the review of implications for management and adaptation. Third, the chapter then considers dimensions of sustainability and the wider lessons of this analysis beyond deltas. As examples, Boxes 11.1, 11.2 and 11.3 synthesise current knowledge for three archetypal deltas in Asia and Africa (see Chapters 2–4) to provide a plausible set of futures to 2050 and beyond.

Box 11.1 Ganges-Brahmaputra-Meghna Delta: Realistic and plausible trends to 2050

Change in economic structure and implications for land use

The GBM Delta will continue to see significant urbanisation with a focus in and around Kolkata and Dhaka, and a corresponding stabilisation or probable decline in rural populations. Greater Dhaka may reach a size bigger than any city existing today. At the same time, the economy will continue to transform and grow rapidly: by 2050 the GDP per capita in the Indian part of the GBM Delta could be seven times larger than today and five times larger in the Bangladeshi part, with major expansion of sectors such as industry and services, and a continued decoupling from agriculture (Chapter 8). Agriculture will decline significantly as a proportion of the economy with intensification of rice (and aquaculture). As a result, labour will leave agriculture in search of higher wages in other sectors. This will be particularly true of smallholders and landless labourers who will find it harder to benefit from the growing service economy and agricultural intensification. This is one key driver of rural to urban migration. Further, agriculture may become more intensive with more mechanisation and larger farm sizes, although no evidence of such trends is yet apparent.

Implications of policies and plans

The Bangladesh Delta Plan 2100 (BDP2100) will trigger more coordinated top-down adaptation and development across Bangladesh, while new infrastructure provision such as the Padma Bridge (scheduled opening 2020) will transform how Bangladesh develops. The Indian portion of the delta would also benefit from its own delta plan, but there is no obvious impetus for this today. The renegotiation of the Ganges Water Treaty from 2026 will have a significant impact on both the Bangladeshi and Indian parts of the delta, though national and state governance frameworks may struggle to adapt to these changes initially.

Biophysical changes and implications for inhabitants and livelihoods

Deltaic accretion is likely to continue in Bangladesh, particularly around the Meghna Estuary, although this may slow, while the Indian part of the GBM Delta will continue to erode (Chapter 5). To 2050, climate change will be a challenge, even with business-as-usual adaptation (Chapter 8). Modelling suggests that wild fisheries are more of a concern, and continued regulation is important to maintain catches. In the short-term (next 10 years) growth in aquaculture looks likely. Beyond this, changes are less certain and over intense aquaculture can lead to abandonment of fish ponds that are difficult to convert to other uses, blighting areas: mixed culture is becoming more popular to overcome this issue. The Sundarbans mangrove forest can persist under expected sea-level rise scenarios to 2100, but has many other threats such as pollution. Beyond 2050, climate change and sea-level rise become a bigger concern, while the expected decline of sediment supply from the rivers will hinder the application of sediment-based and working with nature approaches (Chapters 9 and 10). The wild card of major cyclone landfall remains and this could have severe and disruptive impacts.

Role of adaptation

Under plausible improved adaptation measures, agriculture can continue to prosper and flourish and the delta will remain habitable and productive (Chapter 10). Innovative adaptation approaches such as build elevation via controlled sedimentation (Chapter 5) could be widely implemented across the delta, but this must also address the social issues that it raises. Recent improvements in flood/storm surge warnings and cyclone shelters have greatly improved responses to cyclone, and dikes—if well maintained—are generally considered adequate. The post-2050 challenges require significant preparation which should be integral to today's delta planning. The BDP2100 is arguably transformative and supportive in this regard. It also creates a platform for other transformative adaptation, such as what to do about abandoned fishponds as one example.

11.2 The Anthropocene Transition in Deltas

A key message of the preceding chapters is that deltas are highly dynamic systems in multiple biophysical and socio-economic ways and they are evolving rapidly. Table 11.1 summarises the key observed trends and drivers in deltas. These trends are diverse and linked either directly or indirectly to human activities, consistent with the concept of the Anthropocene. While humans have been influencing deltas for thousands of years in some cases (Bianchi 2016; Welch et al. 2017), these type of changes are now pervasive in low and mid-latitude populated deltas and dominate their evolution. Table 11.2 summarises some potential indicators of the Anthropocene transition in deltas from the analysis discussed in previous chapters.

While within any single discipline the individual trends in Table 11.1 and indicators in Table 11.2 might be unsurprising, the trends are usually seen and managed in isolation. However, in deltas these trends are occurring in the same location and interacting in complex ways. As a result, deltas today are quite different to deltas 30 years ago, reflecting the growing importance of cities, agricultural intensification and the changing structure of the economy. The continuation of human activities means that deltas in 2050 will be different to what is seen today. This evolving inter-relationship of trends contrasts with the common narrative about deltas in a changing world, which often narrowly focuses on just one threat—that of relative sea-level rise—while ignoring the implications of other factors, both singly and in

Table 11.1 Key observed trends in deltas in the late 20th/early 21st century highlighted in this book

Key trends	Drivers	Comments	Chapter
Threatened ecosystems	Human pressures, climate change	Longstanding problem, which is still intensifying, but growing interest in working with nature may indicate a new direction (linked to adaptation)	2, 3, 4
Sediment starvation and relative sea-level rise	Dams, natural subsidence, human-enhanced subsidence, climate-induced sea-level rise	Leads to erosion, loss of elevation and growing flood hazard unless there is significant adaptation	5
Changing risks due to changing hazard, exposure and vulnerability	Growing population and economy, adverse environmental and climate change	Growth in risk is almost inevitable without significant adaptation	6
Growing urban populations/stable or declining and ageing rural populations	Livelihood threats and opportunities, economic/education opportunities, better infrastructure and health facilities in cities, demographic transition, increasing life expectancy	Details depend on fertility, the demographic transition is stronger in Asia than Africa	7

(continued)

Table 11.1 (continued)

Key trends	Drivers	Comments	Chapter
Larger economies: major growth in industry and services and relative decline in agriculture and fisheries	Economic growth, agricultural intensification	Growth in climate and other risks can threaten this trend, so the economic development of deltas is strongly linked to adaptation	8
A growing need for adaptation, especially large scale, planned adaptation	Multiple issues galvanising action such as: climate change, growing numbers of receptors, changing attitudes and a more risk-adverse population	The Bangladesh Delta Plan 2100 illustrates the scale and intervention that is needed: transformative adaptation. What are the best strategic options for such transformation? Retreat, protect, build elevation or whatever? The linkages to catchments and neighbouring seas must also be considered	9
A recognition that adaptation trade-offs and choices have major influence on future delta evolution			10

Table 11.2 Potential indicators of Anthropocene conditions in deltas

Factor	Anthropocene indicators
Sediment supply	Declining trend in sediment supply, with sometimes catastrophic decline compared to pre-Anthropocene conditions, mainly reflecting dam construction
Subsidence	Often enhanced subsidence due to ground fluid withdrawal and drainage, sometimes very rapid (>10 cm/yr) most especially within urban areas on deltas
Land use and land management	Intensification/higher yields to feed the growing delta and associated populations; More diverse, higher value agriculture and aquaculture production (e.g., growth in brackish shrimp aquaculture for export); Move from a subsistence to cash economy and globalisation; Declining agricultural employment
Population/migration	Urbanisation and stable/falling rural populations. Active migration systems, within or adjacent to the delta
Economy	Growing and diversified economy
Natural ecosystems	Decline in biodiversity and natural systems preserved in reserves and protect refuges
Delta-level adaptation	Growing dependence on engineered flood protection (dikes, polders, etc.) as in the Netherlands, leading to lock-in where the choice is between higher defences or major retreat/abandonment; Large-scale integrated and adaptive management up to the whole delta scale (e.g. Bangladesh Delta Plan 2100)

combination. For example, migration from deltas is widely discussed in the climate narrative as being solely a response to sea-level rise, while research summarised in Chapter 7 shows that sea-level rise and environmental degradation more widely is perturbing well-established migration systems, reflecting more general and systemic social trends. Hence sea-level rise and its impacts are only one of many drivers and effects that should be considered when analysing delta evolution and development policies.

The role of adaptation in general and protection through infrastructure in particular is also key to the Anthropocene in deltas. Many deltas could not support their large populations without significant hard adaptation in the form of defence infrastructure. For example, in Bangladesh alone there are more than 6000 km of dikes and embankments around coastal polders and over 2500 cyclone shelters (multi-storey robust buildings) which are a key element in the agricultural and disaster management systems.

These cyclone shelters have been demonstrated to be highly effective to reduce mortality for exposed populations during cyclones (Faruk et al. 2018). They have been transferred to the Mahanadi Delta (Fig. 11.1) where they have similarly been effective (Box 11.2). Significant enhancement of this infrastructure, and its implications for a diverse profile of livelihoods, is ongoing and will continue under the BDP2100 which covers all of Bangladesh (BDP2100 2018). This is discussed more in Sect. 11.3.

11.3 Management and Adaptation of Deltas in the Anthropocene

People living in deltas have long adapted to the changing conditions and situations that characterise the systems, including the seasonal cycles that shape delta life. However, it would seem that the multiple stresses apparent under Anthropocene conditions mean that historic approaches to adaptation and management will not be enough in the future. This mandates an intensification and transformation

Fig. 11.1 Effective local adaptation infrastructure: Cyclone shelter in the Mahanadi Delta, India (Photo: Amit Ghosh/Shouvik Das)

of adaptation in terms of technical approach, scale and governance arrangements.

The emergence of delta-scale plans (Seijger et al. 2017) would seem to be one appropriate response to modifying the approach to adaptation in deltas. Being in many ways a response to the flooding of New Orleans in Hurricane Katrina in 2005, these started in the Netherlands with the Delta Plan in 2008 (Deltacommissie 2008; Kabat et al. 2009; Stive et al. 2011). This approach to delta planning has been exported to the Mekong Delta, Vietnam (Mekong Delta Consortium 2013) and Bangladesh (BDP2100 2018), with other deltas considering similar plans. The Mississippi Delta also has a major delta planning process (CPRAL 2017).

All these deltas plans and planning processes stress the importance of governance and integration and a long-term commitment to adaptation. For example, the Netherlands has created a Delta Commission and a Delta Commissioner role with long-term (20 year) financial commitments and new institutions/governance approaches to manage the delta; a structure which has the potential to be translated to other deltas. However, it is important not to see delta governance approaches in isolation, as they are heavily influenced and directed by governance frameworks across multiple administrative and hydrological scales. The Dutch system, for example, operates within the confines of the river basin management planning framework under the Water Framework, Floods and Environmental Impact Assessment Directives (among others), but also under international legal arrangements for the Meuse, Rhine and Scheldt Rivers. Likewise, the implementation of the Mekong Delta Plan is dependent to some extent on the Mekong River Convention and on relevant (though not delta-specific) national policy and legal contexts.

Such delta plans promote a systems view of adaptation which allows important changes in adaptation and management to be incorporated. For example, there is growing interest in soft infrastructure and "building with nature" approaches (van Wesenbeeck et al. 2014) and hybrid approaches that combine soft and hard adaptation (Smajgl et al. 2015); these can include maintaining or recreating mangrove and forest belts and controlled sedimentation within polders to raise land levels. Natural system approaches are most strongly developed in the Mississippi Delta, USA (Costanza et al. 2006; Day et al. 2007; Paola et al. 2010), reflecting a desire to counter massive historic losses of wetland since the 1950s. Further innovation in delta adaptation is anticipated over the next few decades and working with nature to the maximum degree possible, especially for sediment management, is one guiding principle of these efforts.

An important issue when considering future adaptation is the notion of "lock-in". Lock-in occurs when any decisions limit or curtail future options. Building embankments and dikes with polders in a delta with relative sea-level rise, and growing economic value behind the defences leads to choice between raising these defences or a major abandonment/retreat. This dilemma epitomises much of today's Netherlands and other wealthy developed countries. Thus, defences can lock the delta into a pathway that is unsustainable in the long-term and can only be reversed with great efforts and significant resources. Seijger et al. (2018) conceptualise that the delta pathway lock-in can occur for technological and institutional reasons that co-evolve. Skills, relations and interests of an institution can limit their willingness to take up a new technology. On the other hand, if a technology is widely deployed, it is easier and cheaper to maintain and improve than replace with a completely new approach affecting assets and way of living. This is what Seijger et al. (2018) call the dual lock-in of deltas. However, environmental concerns of societies, especially in wealthy countries, can also have a bearing on delta development and planning (van Staveren and van Tatenhove 2016; Welch et al. 2017). In such cases, the competing interests between hazard exposure, economic development, social welfare and environmental protection make delta planning more complex, and a clear understanding of trends, threats and trade-offs is essential (Suckall et al. 2018; Chapter 9).

As an alternative adaptation pathway, controlled sedimentation and building elevation may create a pathway where the land surface can keep relative pace with sea level—the sediment that allows this strategy has been characterised as "brown gold" (Darby et al., 2018). This is a more sustainable adaptation strategy in deltas if sufficient sediment is available. However, to be sustainable, land raising will need to be an ongoing process and the effects on society during the time that the land is in the process of being raised needs to be considered; this requires a strategic approach. More broadly, questions of equity and how the poorest and politically-underrepresented are treated in such processes, especially the possibility of being displaced from their land, must be addressed.

As noted in Boxes 11.1–11.3, the commitment and interest in delta planning in the Mahanadi, Volta and Indian portion of the GBM Delta is currently lower than in Bangladesh. Would they benefit from their own delta plan? Or should they be managed together with the neighbouring non-deltaic coasts? In the Mahanadi context, there may be good reasons for the development of a delta plan by the state of Odisha, if only in order to inform the current controversy over water use in the upstream state of Chhattisgarh and where, although legal frameworks on coastal zone management are actively enforced, broader water resource management legislation across the basin is much less effective.

In other cases, this question needs to be explored further. As part of these discussions, two additional questions also need to be considered: how should this management be conducted, and what is a sustainable delta?

Box 11.2 Mahanadi Delta: Realistic and plausible trends to 2050

Change in economic structure and implications for land use

The Mahanadi Delta will see many trends similar to the GBM delta. There will be significant urbanisation with a particular focus in and around Bhubaneswar, and a continued decline in rural populations (Chapter 7). However, the scale of these cities is small compared to Dhaka and Kolkata in the GBM Delta. Again, the economy will continue to transform and grow rapidly: by 2050 the GDP per capita could be 5.4 times larger than today, with major expansion of sectors such as industry and services, and a decline in the relative size and employment in agriculture (Chapter 8).

Implications of policies and plans

Separate delta level efforts need to be initiated to counter the decline of forest and biodiversity and combat increasing pollution load in the ecosystem. In governance terms, increasing conflict between the Indian states of Chhattisgarh and Odisha may prompt the determination of state water allocations, but the delta is unlikely to feature strongly in this process and may suffer from increasing water scarcity in future. This creates little impetus for the development of strategic delta planning (Chapter 9).

Biophysical changes and implications for inhabitants and livelihoods

Erosion is already widespread in the delta and this is likely to continue and may trigger hard or soft (working with nature) engineering responses (Chapter 5). The wild card of a major cyclone landfall remains a concern and this could have severe impacts as in the 1999 Super Cyclone. Beyond 2050, climate change and sea-level rise become a bigger concern, and sediment supply from the rivers is expected to decline, hindering working with nature approaches (Chapter 5). These upcoming challenges require preparation now and should be integral to delta planning today, but the institutional framework and willingness to facilitate this do not exist.

Role of adaptation

Adaptation by individual households is likely to continue as is, without a larger development plan for the delta. This is likely to continue to be driven by the need to reduce vulnerability, e.g. migration, loans and improving individual homes. Hence, the emergence of women headed household as a separate vulnerable group within the delta is likely to persist. While this may improve the opportunities for those able to afford adaptation, those unable to find the resources to adapt are likely to remain in poverty and fall further behind. However, recent improvements in disaster risk management, including flood warnings, evacuations and the provision of cyclone shelters have greatly improved responses—Cyclone Phailin in 2013, while causing widespread damage, had hardly any casualties. Such planned delta level adaptation would be increasingly useful beyond 2050.

11.4 Deltas and Sustainability

The idea of sustainability is key to the concept of the Anthropocene. The analysis in this book shows that sustainability has multiple dimensions and timescales. Day et al. (2016) consider delta sustainability within the context of global biophysical and socio-economic constraints, recognising geomorphic, ecological and economic aspects of delta sustainability. However, a narrow focus on the physical processes that underpin delta functioning tends to underemphasise the influence of maintaining a sustainable delta society, which includes livelihood sustainability, demography and well-being. The following therefore highlight trade-offs inherent in the concept of adaptation when considered in relation to the different aspects of delta sustainability, focussing on the examples raised in the detailed delta studies.

Geomorphic Sustainability

Geomorphic sustainability links sediment budgets and resulting land elevation (see Chapter 5). Sediment flux to the deltas from the catchment is often significantly reduced by upstream dams, but even within deltas, sediment movement is widely restricted by engineering interventions such as dikes for urbanisation and flood defence. Subsidence in deltas is a naturally occurring process due to sediment compaction (Meckel et al. 2007; Syvitski 2008) that can be greatly accelerated by groundwater abstraction (e.g. for irrigation or for drinking water). The mean subsidence of 46 major deltas is 3.6 mm/yr, but can reach at least 22 mm/yr in extreme cases such as the Indus Delta in Pakistan (Tessler et al. 2018), and even more in urban areas on deltas (Nicholls 2018). This, combined with one metre or more sea-level rise, will result in significant land areas in deltas being below sea level by the end of this century (Syvitski et al. 2009; Brown et al. 2018); land which will either be submerged or dependent on major defences and drainage systems (as in the Netherlands) (Fig. 11.2).

The benefits of regular flooding and sedimentation for deltas is highlighted by Auerbach et al. (2015). They estimated that poldered areas in coastal Bangladesh have lost 1.0–1.5 m land elevation since the 1960s compared to the neighbouring Sundarbans mangrove forest whose land elevation has remained stable relative to sea level. However, if such subsided lands are reconnected with tidal inundation as a result of a natural disaster, or through controlled flooding, rapid sedimentation of around tens of centimetres of elevation increase in months can occur (Auerbach et al. 2015). There are examples of small-scale controlled inundation practices to locally raise land levels, such as the Tidal River

Fig. 11.2 Erosion and flooding are both indicative of sediment supply and sea-level rise issues (Photo: Mousuni Island, GBM Delta—Shouvik Das)

Management (TRM) in Bangladesh (Chapter 2; Box 11.1), aiming to ensure long-term geomorphic and ecological sustainability. Even though such interventions provide long-term benefits of reduced flooding and waterlogging, TRM remains controversial as it results in the temporary loss of productive land and hence livelihoods. Without addressing the short-term institutional limitations and providing compensation, many communities are therefore reluctant to implement it (Gain et al. 2017). Future plans for TRM are consequently limited, and it is not included within more than 100 proposals contained within the BDP2100. A more ambitious plan in Bangladesh could aim to raise larger areas of land and try to keep pace with relative sea-level rise, but a change in mindset will be required.

Eliminating the sediment retention of upstream dams is similarly problematic as many of these dams are located in a different country, and they serve multiple economic purposes such as providing irrigation water, producing electricity and ensuring navigation. Even if sediment can bypass the dam, the regulation of flow greatly reduces downstream sediment transport to the delta. Thus, competing economic interests often cause sediment starvation of the coastal areas even without the negative effect of embankments and water abstraction. The widespread erosion of the Volta Delta due to sediment starvation from upstream dams is well known (Chapter 4).

As noted earlier, long-term adaptation and planning in deltas are becoming more widespread around the world, aiming to balance the geomorphic and societal needs of the delta (MDP 2013; BDP2100 2018; DP 2018). However, there are ultimately three policy choices in the face of the long-term (i.e. 2100 and beyond) sea-level rise and subsidence: (i) abandon the coastal zone; (ii) protect the population of the coastal zone with ever-higher defences, including pumping; or (iii) build land elevation by controlled sedimentation (Nicholls et al. 2018a). Assuming climate stabilisation and sufficient sediment supply can be maintained from upstream catchments, innovative solutions for sediment management are proposed, but their long-term feasibility and social acceptability need further trials (Day et al. 2014; Gain et al. 2017; also Chapter 5).

Ecological Sustainability

The ecology of populated deltas is highly modified and agriculture and aquaculture have largely replaced natural systems, except for protected areas like the Sundarbans. Day et al. (2016) argued that river inputs regulate soil processes thus enabling better accretion rates, soil formation and resistance to erosion and sea-level rise. But ecological sustainability should be viewed more broadly. Saline intrusion in rivers and groundwater results in soil salinisation (Fan et al. 2012) and the degradation of aquatic and terrestrial biodiversity (Goss 2003). Salinisation also requires radically altered agriculture practices (Rahman et al. 2011; Renaud et al. 2015). However, soil salinisation and biodiversity degradation is also driven by economic transformations of land use that can be highly detrimental to the agricultural yields and may encourage maladaptive processes (Fig. 11.3). An example of such a maladaptation, from a sustainability point of view, is the large-scale

Fig. 11.3 Extensive shrimp farming replaces traditional agricultural practices and local ecosystems in the GBM Delta (Photo: Attila Lázár)

high intensity brackish shrimp cultivation in Bangladesh, which increases soil salinity, acidity, degrades soil quality and can also result in soil toxicity and mangrove destruction (Ali 2006; Azad et al. 2009). While unplanned and uncoordinated shrimp cultivation creates short term export value, it also creates social conflicts, as it negatively impacts traditional agriculture and has significant negative ecological consequences (Flaherty et al. 1999; Hossain et al. 2013; Paul and Røskaft 2013). For example, the collection of shrimp larvae is an important informal livelihood, and one that is accessible for women (Ahmed et al. 2010), but the systematic removal of larvae from coastal waters has significant negative impacts on ecosystem services and aquatic ecology (Hoq 2007; Azad et al. 2009). Also, even though shrimp cultivation is lucrative for the private interests involved, the realised benefits of the shrimp production to delta residents is small due to the outflow of profit to the investors (Swapan and Gavin 2011). Virtually all shrimp produced in the GBM Delta in Bangladesh is exported (Quassem et al. 2003), and hence does not support local food security and availability.

In natural areas such as the Sundarbans—the world's largest mangrove forest (Chapter 2)—major land losses are widely expected due to sea-level rise. However, there is good evidence that these systems are more resilience and can persist, even under high rates of sea-level rise and subsidence, although this is dependent on the availability of sediment and does not mean that the current species composition is conserved. In the Sundarbans, there is a change to more salt-tolerant mangroves (Payo et al. 2016; Mukhopadhyay et al. 2018). Other non-climate risks are noteworthy such as oil spills and pollution—the Sundarbans contain major shipping routes.

Thus, ecological sustainability partially depends on geomorphological processes, but can also be heavily impacted by human activities in the short-term. In the long-run, ecological sustainability must be linked to geomorphic sustainability.

Economic Sustainability

From the economic perspective, the future sustainability of deltas will depend on their capacity to provide the goods and services required to ensure the well-being of their population and contribute to the achievement of the Sustainable Development Goals (SDGs) (Nicholls et al. 2018b). But there are significant potential trade-offs involved when some elements of the sustainability are prioritised. For example, shrimp farming is very lucrative and contributes to economic growth (SDG8), being the fifth largest export commodity of Bangladesh at about US$500 million in 2016–2017 (BBS 2017) while simultaneously damaging water quality and affecting biodiversity conservation (SGDG15). Similarly, coastal protection protects economic activities from extreme events, but this threatens geomorphic sustainability.

Economic sustainability can, in theory, be steered through policies, research and development and investment. However, stark dilemmas exist between seeking the benefits of industrialisation and protection of the environment, biodiversity and the traditional way of living. The sustainability dilemmas highlight issues of reversibility, the potential for technology, and the inertia in policy focussed on economic growth. A planned Indian-Bangladeshi cross-border coal power plant, for example, endangers the Sundarbans mangrove forest and the wider region through water and air pollution (Ghosh 2018). Yet India is rapidly moving away from coal power as a redundant resource, seeking to implement solar and other renewable energy technologies that are less polluting for climate and local environments (Mehra and Bhattacharya 2019).

The drive for economic growth is acute in delta regions, because of their role as growth poles of population and economic activity in virtually all deltas (Chapter 8). In effect, natural capital in deltas is being converted to physical and financial capital. Principles of sustainability suggest that such a strategy can be sustained as long as natural capital thresholds are not exceeded. Without knowledge of where many thresholds and tipping points for natural systems are in deltas, there is a significant priority to conserve natural systems and processes to keep future development options open and to ensure the avoidance of overall loss of sustainability potential.

Social Sustainability

Economic sustainability does not necessarily mean that the delta can equitably provide livelihoods and well-being. An increasing GDP or an increase in average living standards does not mean that poverty is reduced for everyone (Ravallion 2001). The poor typically benefit from economic growth, but they are also disproportionally hurt from economic contraction or hazard events. Indeed, increased economic growth can drive inequality, where poorer sectors of society are left out of growth or are driven from areas of development as land values rise (Amoako-Johnson et al. 2016). Unequal benefits of farming technologies are also highlighted by Chapman and Darby (2016) in the Mekong Delta, Vietnam. They showed that the current triple-cropping strategy with protection from dikes only benefits the wealthier farmers with a limited 10-year maximum benefit window as soil nutrients are depleted. However, taking a more sustainable view and leave the sluice gates open, agriculture could become more equitable although the net economic benefits would become smaller.

If rural delta regions do not prioritise liveability and livelihood security, such regions will inevitably lose populations to cities. Environmental dimensions of migration in deltas include land grabbing for agricultural intensification, land degradation and marginalised livelihoods. These exacerbate existing migration trends relating to family obligations, seeking better education and infrastructure (Chapter 7). In addition, exposure to climate shocks often displaces populations temporarily but changes the long term attractiveness of permanent moves, almost invariably to urban settlements (Fig. 11.4). Rural coastal communities are exposed to environmental stress and hazards, and often lack education and health infrastructure that can be a serious motivation for migration. Societal sustainability therefore incorporates the spatial distribution of populations, the links between cities and rural hinterlands and liveable environments where inequality and poverty and minimised.

Deltas and the Sustainable Development Goals

The 17 SDGs express aspirations for human development that require the realisation of a universal, but diverse set of ethical principles, such

Fig. 11.4 Displacement of populations by floods can lead to permanent migration (Photo: A. K. M. Saiful Islam)

as inclusion, justice, equality, dignity, well-being, global solidarity, sharing, sustainability and public participation (United Nations 2015; Szabo et al. 2016; Hutton et al. 2018). As with all coupled human-environment systems, the issues that impact deltas in the Anthropocene are both diverse and complex (Young et al. 2006). It is for this reason that the SDGs are extensive and comprehensive in their approach to capturing the biophysical and socio-economic context for development, as well as the inherent trade-offs associated with this development (Hutton et al. 2018). Economic growth, poverty, environmental degradation, and inequality as well as food production are recurrent issues. As such, of particular interest within delta systems is the relationship between the SDG goals in Table 11.3, although all the SDGs are linked.

These goals can, and do, directly compete with each other through the sustainability of food production processes and the demands of urbanisation (Machingura and Lally 2017). Decision makers are

Table 11.3 Selected SDG goals of particular relevance in the future management of deltas

SDG	Aim
1	(No poverty) Eradicating poverty in all its forms by 2030
2	(No hunger) End all forms of hunger and malnutrition by 2030
8	(Decent work and economic growth) Promote sustained economic growth and higher levels of productivity
10	(Reduced inequalities) Reduce inequality within and among countries
14	(Life below water) To sustainably manage and protect marine and coastal ecosystems
15	(Life on land) To conserve and restore the use of terrestrial ecosystems

therefore faced with choices regarding the intensification of agriculture that, while enhancing food production, may also reduce the demand for labour, increase inequalities, undermine subsistence and small holding livelihoods as well as causing damage to the terrestrial and aquatic ecosystems. Similarly, the development of infrastructure, so important for reducing rural poverty, can also lead to highly destructive environmental practices. Such insight is extremely relevant for policy development, as it calls for processes of trade-off decisions, compromise and planning of strategic development pathways. The recognition of this raises critical ethical questions about potential pathways and compromises to achieve a balance between the SDGs, and demands transparent scrutiny of priorities and motivations for sustainable development, as well as the identification of winners and losers (Hutton et al., 2018).

Box 11.3 Volta Delta: Realistic and plausible trends to 2050

Biophysical changes and implications for inhabitants and livelihoods

The Volta Delta differs from the GBM and Mahanadi Deltas in that the main marine hazard is erosion and flooding at the immediate coast—tides and surges are small and penetrate relatively small distances inland compared to the other deltas. Other delta processes have been removed for 50 years or more by upstream dams so river floods and a new sediment supply are almost totally absent (Chapter 5). These trends are likely to continue and accelerate, as are the engineered protection responses of breakwaters, groynes and nourishment in developed areas.

This protection is certainly buying time in the more critical locations on the open coast like Keta, but could exacerbate erosion to the east, causing potential conflict with Togo. Beyond 2050, climate change and sea-level rise become a bigger concern for the delta with a growing flood plain and the potential for more widespread impacts.

Change in economic structure and implications for land use
There will be significant rural to urban outmigration to Accra, Tema, Lomé (Togo) and other urban areas mainly due to livelihood decisions (Chapter 7). Looking to 2050 and beyond, the urban footprint of Greater Accra and Greater Lomé may significantly expand on to the Volta Delta. As the total fertility rate is higher in West Africa than in Asia, the rural population may be stable or even rise rather than decline as in the Asian deltas. This in turn is likely to drive land cover change towards agriculture and more intense agriculture, although urbanisation may become a feature in a few decades as the neighbouring cities grow. Similarly aquaculture on the Keta Lagoon is expected to intensify greatly. Inland, climate variability and change may be manifest by drought and its impact on agriculture, with the impacts beyond the delta (Chapter 6). As with the other deltas, the delta economy will continue to transform and grow rapidly: by 2050 the GDP per capita in the delta could be 4.5 times larger than today, with major expansion of sectors such as industry and services, and a decline in the relative size and employment in agriculture (Chapter 8).

Implications of policies and plans
In governance terms, the specific needs of the delta are not well recognised as a distinct feature, but the erosion of the coast is widely recognised as a problem, as demonstrated by the government-funded adaptation projects. As with the Mahanadi, this creates little impetus for the development of strategic delta planning. Upcoming challenges require preparation now and should be integral to delta planning today, but the institutions for this to happen do not seem to exist. In the absence of delta planning, climate change is likely to increase migration away from the delta.

Role of adaptation
While more adaptation is to be expected due to ongoing erosion, climate and other changes, this is likely to be more of the same (Chapter 9). A key innovation would be the development of more integrated beach and erosion management and recognising the value of beach sand. With the absence of cyclones and local storms, erosion will continue to displace people where there is no protection, but this will be small compared to other drivers of migration. However, the images of these impacts are evocative and will continue to draw significant attention.

11.5 Insights on the Anthropocene Transition and Its Management

Beyond deltas, these analyses provide an important and useful perspective on the Anthropocene in general that are applicable for other coupled human-environment systems. In deltas, geomorphic, ecological, economic and social processes coexist in the same space and interact in multiple ways. They show that multiple drivers are in operation today and these are linked, directly or indirectly, to human agency. Importantly, this includes adaptation that, while a response to these drivers, also potentially has an important feedback on the future evolution of deltas (Welch et al. 2017; Seijger et al. 2018). This brings out the complexity of these systems and their evolution is difficult to predict with confidence—different possible trajectories are apparent and policy can steer change towards desirable outcomes. Similar processes are operating in other coupled human-environment systems, such as river catchments (upstream of the deltas considered here), drylands or coastal lowlands. In all cases, this suggests that a systems analysis is useful and instructive to analyse the current status of these areas, and to consider possible future trajectories and inform policy.

It is also important to recognise the complementary role of bottom-up and top-down approaches in such analyses (Conway et al. 2019). Climate science often takes a "global" view, but this can miss important social and economic changes that manifest at more local levels. Integrating these approaches leads to a more complete analysis and recognition of both the broad-scale and more local drivers of change and sets the global driver of climate change (and other broad-scale drivers) into an appropriate context. This means engaging with stakeholders in a participatory manner is a key component of success in managing and developing coupled human-environment systems in the Anthropocene (e.g. Nicholls et al. 2018b).

11.6 Key Lessons

This book identifies a number of important dimensions of sustainability for deltas in the Anthropocene. Firstly, it demonstrates that deltas are hotspots in the Anthropocene transition and exemplify many of the diverse social and environmental changes that are occurring across the planet. The range of biophysical and socio-economic processes interacting in deltas, including migration flows and adaptation choices, are shaping their future development and sustainability or persistence. Deltas are, therefore, about more than sediment supply and geomorphic change, but sediment availability and sedimentation emerge as key factors influencing long-term delta survivability. Over the next 30 years, all the deltas considered here have potential to develop in ways that will deliver great benefits for their populations (Boxes 11.1–11.3). Economic development can create significant adaptive capacity to address the challenges that lie beyond this time frame, especially if the framework for such responses is planned now rather than delayed.

Strategic adaptation and consideration of delta development trajectories can help to avoid lock-in and other irreversible decisions—strategic raising of agricultural and natural areas with controlled sedimentation should be encouraged and become the norm where possible. However, it is unclear if sufficient sediment will be available in all cases, and this does not address urban areas where more traditional flood defence is likely to remain the key form of risk management. The prospect of occasional large floods in delta cities is a significant concern (cf. Hallegatte et al. 2013). This raises the question about the trade-off between elevation and wealth. For example, Day et al. (2016) and many other sources see elevation in deltas as fundamental. Yet the Netherlands presently copes with such conditions of lost elevation, reflecting its high wealth, access to technology and good governance. Hence, while the long term health of deltas is dependent on sediment supply, wealthy societies in deltas can, in effect, buy time and adapt with technology, and economic growth greatly increases the capacity of these options.

Long-term adaptation and planning is becoming more common in deltas around the world aiming to balance the biophysical and societal needs. However, as we look to the long-term (i.e. 2100 and beyond), there are ultimately three policy choices for deltas in the face of sea-level rise and subsidence: (i) abandon the coastal zone; (ii) protect the population of the coastal zone with ever-higher defences, including pumping, and consideration of residual risk; or (iii) raise land elevation by controlled sedimentation. As noted, this dilemma can be dodged in the short-term, but as time goes on this stark choice will become clearer and clearer. In populated deltas, building elevation is an attractive option if sufficient sediment is available and it can be delivered to the delta surface in a manner that is socially acceptable. The absence of such proposals in the current BDP2100 is noteworthy and much work remains to be done in promoting this approach. However, the BDP2100 will be regularly updated allowing such innovation. In contrast, in the Volta Delta controlled sedimentation does not seem to be an option as there is no sediment supply and other approaches are required.

Many earlier analyses paint a negative picture of the future of deltas and suggest that, without radical transformation, disaster lies ahead (e.g. Syvitski et al. 2009; Day et al. 2016). This book recognises those major challenges, but it also highlights opportunities and possible trajectories where these deltas will prosper, in some senses the potential for a good Anthropocene (Bennett et al. 2016). These futures depend on significant adaptation, be it working with nature to build elevation via controlled sedimentation, or more traditional hard adaptation approaches such as building and improving dikes and embankments and moving towards a situation resembling the Netherlands. Bennett et al. (2016) define "seeds" of a good Anthropocene, and in deltas this would seem to include working with natural approaches to maximise geomorphic sustainability. Importantly, delta scale simulations, including human agency, are becoming feasible (Angamuthu et al. 2018), facilitating this approach. Equally, there are trajectories where collapse may occur especially under scenarios of degraded livelihoods and extreme events. To a great extent, the future effectiveness of delta-specific approaches will be tempered by the overall quality of governance in basins as a whole. It is important to recognise that not all deltas will behave or respond in the same ways, and success stories in some deltas may be offset by failure in others.

References

Ahmed, N., Troell, M., Allison, E. H., & Muir, J. F. (2010). Prawn postlarvae fishing in coastal Bangladesh: Challenges for sustainable livelihoods. *Marine Policy, 34*(2), 218–227. https://doi.org/10.1016/j.marpol.2009.06.008.

Ali, A. M. S. (2006). Rice to shrimp: Land use/land cover changes and soil degradation in southwestern Bangladesh. *Land Use Policy, 23*(4), 421–435. https://doi.org/10.1016/j.landusepol.2005.02.001.

Amoako-Johnson, F., Hutton, C. W., Hornby, D., Lázár, A. N., & Mukhopadhyay, A. (2016). Is shrimp farming a successful adaptation to salinity intrusion? A geospatial associative analysis of poverty in the populous Ganges–Brahmaputra–Meghna Delta of Bangladesh. *Sustainability Science, 11*(3), 423–439. https://doi.org/10.1007/s11625-016-0356-6.

Angamuthu, B., Darby, S. E., & Nicholls, R. J. (2018). Impacts of natural and human drivers on the multi-decadal morphological evolution of tidally-influenced deltas. *Proceedings of the Royal Society A: Mathematical, Physical and Engineering Science, 474,* 2219. https://doi.org/10.1098/rspa.2018.0396.

Auerbach, L. W., Goodbred, S. L., Jr., Mondal, D. R., Wilson, C. A., Ahmed, K. R., Roy, K., et al. (2015). Flood risk of natural and embanked landscapes on the Ganges-Brahmaputra tidal delta plain. *Nature Climate Change, 5,* 153–157. https://doi.org/10.1038/nclimate2472.

Azad, A. K., Jensen, K. R., & Lin, C. K. (2009). Coastal aquaculture development in Bangladesh: Unsustainable and sustainable experiences. *Environmental Management, 44*(4), 800–809. https://doi.org/10.1007/s00267-009-9356-y.

BBS. (2017). *Bangladesh statistics 2017.* Bangladesh Bureau of Statistics, Statistics and Informatics Division (SID), Ministry of Planning, Bangladesh. http://bbs.portal.gov.bd/sites/default/files/files/bbs.portal.gov.bd/page/a1d32f13_8553_44f1_92e6_8ff80a4ff82e/Bangladesh%20%20Statistics-2017.pdf.

BDP2100. (2018). *Bangladesh delta plan 2100. Volumes 1-Strategy and 2-Investment plan.* General Economics Division (GED), Bangladesh Planning Commission, Government of the People's Republic of Bangladesh, Dhaka, Bangladesh. https://www.bangladeshdeltaplan2100.org/. Last accessed 8 October 2018.

Bennett, E. M., Solan, M., Biggs, R., McPhearson, T., Norström, A. V., Olsson, P., et al. (2016). Bright spots: Seeds of a good Anthropocene. *Frontiers in Ecology and the Environment, 14*(8), 441–448. https://doi.org/10.1002/fee.1309.

Bianchi, T. S. (2016). *Deltas and humans: A long relationship now threatened by global change*. Oxford, UK: Oxford University Press.

Brown, S., Nicholls, R. J., Lázár, A. N., Hornby, D. D., Hill, C., Hazra, S., et al. (2018). What are the implications of sea-level rise for a 1.5, 2 and 3 °C rise in global mean temperatures in the Ganges-Brahmaputra-Meghna and other vulnerable deltas? *Regional Environmental Change, 18*(6), 1829–1842. http://dx.doi.org/10.1007/s10113-018-1311-0.

Chapman, A., & Darby, S. (2016). Evaluating sustainable adaptation strategies for vulnerable mega-deltas using system dynamics modelling: Rice agriculture in the Mekong Delta's An Giang Province, Vietnam. *Science of the Total Environment, 559*, 326–338. https://doi.org/10.1016/j.scitotenv.2016.02.162.

Conway, D., Nicholls, R. J., Brown, S., Tebboth, M., Adger, W. N., Bashir, A., et al. (2019). The need for bottom-up assessments of climate risks and adaptation in climate-sensitive regions. *Nature Climate Change, 9*, 503–511. https://doi.org/10.1038/s41558-019-0502-0.

Costanza, R., Mitsch, W. J., & Day, J. W., Jr. (2006). A new vision for New Orleans and the Mississippi Delta: Applying ecological economics and ecological engineering. *Frontiers in Ecology and the Environment, 4*(9), 465–472. http://dx.doi.org/10.1890/1540-9295(2006)4[465:ANVFNO]2.0.CO;2.

CPRAL. (2017). *Louisiana's comprehensive master plan for a sustainable coast: Commited to our coast*. Baton Rouge, LA: Coastal Protection and Restoration Authority of Louisiana. http://coastal.la.gov/wp-content/uploads/2017/04/2017-Coastal-Master-Plan_Web-Single-Page_CFinal-with-Effective-Date-06092017.pdf. Last accessed 20 December 2018.

Darby, S. E., Nicholls, R. J., Rahman, M. M., Brown, S., & Karim, R. (2018). A sustainable future supply of fluvial sediment for the Ganges-Brahmaputra Delta. In R. J. Nicholls, C. W. Hutton, W. N. Adger, S. E. Hanson, M. M. Rahman, & M. Salehin (Eds.), *Ecosystem services for well-being in deltas: Integrated assessment for policy analysis* (pp. 277–291). Cham: Springer. http://dx.doi.org/10.1007/978-3-319-71093-8_15.

Day, J. W., Agboola, J., Chen, Z., D'Elia, C., Forbes, D. L., Giosan, L., et al. (2016). Approaches to defining deltaic sustainability in the 21st century. *Estuarine, Coastal and Shelf Science, 183*, 275–291. http://dx.doi.org/10.1016/j.ecss.2016.06.018.

Day, J. W., Boesch, D. F., Clairain, E. J., Kemp, G. P., Laska, S. B., Mitsch, W. J., et al. (2007). Restoration of the Mississippi Delta: Lessons from hurricanes Katrina and Rita. *Science, 315*(5819), 1679. http://dx.doi.org/10.1126/science.1137030.

Day, J. W., Kemp, G. P., Freeman, A., & Muth, D. P. (2014). *Perspectives on the restoration of the Mississippi Delta: The once and future delta*. Netherlands: Springer.

Deltacommissie. (2008). Working together with water. A living land builds for its future. *Findings of the Delta Commissie*. The Netherlands: Delta Commissie. http://www.deltacommissie.com/doc/deltareport_full.pdf. Last accessed 28 August 2017.

DP. (2018). *Delta programme 2018—Continuing the work on a sustainable and safe delta*. The Netherlands: Delta programme. https://deltaprogramma2018.deltacommissaris.nl/viewer/publication/1/2-delta-programme.

Fan, X., Pedroli, B., Liu, G., Liu, Q., Liu, H., & Shu, L. (2012). Soil salinity development in the yellow river delta in relation to groundwater dynamics. *Land Degradation & Development, 23*(2), 175–189. http://dx.doi.org/10.1002/ldr.1071.

Faruk, M., Ashraf, S. A., & Ferdaus, M. (2018). An analysis of inclusiveness and accessibility of cyclone shelters, Bangladesh. *Procedia Engineering, 212*, 1099–1106. https://doi.org/10.1016/j.proeng.2018.01.142.

Flaherty, M., Vandergeest, P., & Miller, P. (1999). Rice paddy or shrimp pond: Tough decisions in rural Thailand. *World Development, 27*(12), 2045–2060. https://doi.org/10.1016/S0305-750X(99)00100-X.

Gain, A. K., Benson, D., Rahman, R., Datta, D. K., & Rouillard, J. J. (2017). Tidal river management in the south west Ganges-Brahmaputra Delta in Bangladesh: Moving towards a transdisciplinary approach? *Environmental Science & Policy, 75*, 111–120. https://doi.org/10.1016/j.envsci.2017.05.020.

Ghosh, S. (2018). A cross-border coal power plant could put Sundarbans at risk. *The Wire*.

Giosan, L., Syvitksi, J. P. M., Constantinescu, S., & Day, J. (2014). Climate change: Protect the world's deltas. *Nature, 516*, 31–33. https://doi.org/10.1038/516031a.

Goss, K. F. (2003). Environmental flows, river salinity and biodiversity conservation: Managing trade-offs in the Murray–Darling basin. *Australian Journal of Botany, 51*(6), 619–625. https://doi.org/10.1071/BT03003.

Hallegatte, S., Green, C., Nicholls, R. J., & Corfee-Morlot, J. (2013). Future flood losses in major coastal cities. *Nature Climate Change, 3*, 802. http://dx.doi.org/10.1038/nclimate1979.

Hoq, M. E. (2007). An analysis of fisheries exploitation and management practices in Sundarbans mangrove ecosystem, Bangladesh. *Ocean*

& Coastal Management, 50(5), 411–427. https://doi.org/10.1016/j.ocecoaman.2006.11.001.

Hossain, M. S., Uddin, M. J., & Fakhruddin, A. N. M. (2013). Impacts of shrimp farming on the coastal environment of Bangladesh and approach for management. *Reviews in Environmental Science and Bio/Technology, 12*(3), 313–332. https://doi.org/10.1007/s11157-013-9311-5.

Hutton, C. W., Nicholls, R. J., Lázár, A. N., Chapman, A., Schaafsma, M., & Salehin, M. (2018). Potential trade-offs between the Sustainable Development Goals in coastal Bangladesh. *Sustainability, 10*(4), 1008. http://dx.doi.org/10.3390/su10041108.

Kabat, P., Fresco, L. O., Stive, M. J. F., Veerman, C. P., van Alphen, J. S. L. J., Parmet, B. W. A. H., et al. (2009). Dutch coasts in transition. *Nature Geoscience, 2*, 450. http://dx.doi.org/10.1038/ngeo572.

Machingura, F., & Lally, S. (2017). *The Sustainable Development Goals and their trade-offs* (ODI Development Progress, Case Study Report). London, UK: Overseas Development Institute (ODI). https://www.odi.org/publications/10726-sustainable-development-goals-and-their-trade-offs. Last accessed 4 April 2019.

MDP. (2013). *Mekong Delta plan: Long-term vision and strategy for a safe, prosperous and sustainable delta* (pp. 126 p.). https://www.wur.nl/upload_mm/2/c/3/b5f2e669-cb48-4ed7-afb6-682f5216fe7d_mekong.pdf.

Meckel, T. A., Ten Brink, U. S., & Williams, S. J. (2007). Sediment compaction rates and subsidence in deltaic plains: Numerical constraints and stratigraphic influences. *Basin Research, 19*(1), 19–31. https://doi.org/10.1111/j.1365-2117.2006.00310.x.

Mehra, M., & Bhattacharya, G. (2019). Energy transitions in India: Implications for energy access, greener energy, and energy security. *Georgetown Journal of Asian Affairs*. http://hdl.handle.net/10822/1053156.

Mekong Delta Consortium. (2013). *Mekong Delta plan: Long-term vision and strategy for a safe, prosperous and sustainable delta*. Partners for Water, Netherlands. Ministry of Infrastructure and Environment, Embassy of the Kingdom of the Netherlands, Hanoi Ministry of Natural Resources and Environment, Ministry of Agriculture and Rural Development Consortium, Royal HaskoningDHV, WUR, Deltares, Amersfoort, Netherlands. https://www.mekongdeltaplan.com/storage/files/files/mekong-delta-plan.pdf?1. Last accessed 10 April 2019.

Mukhopadhyay, A., Payo, A., Chanda, A., Ghosh, T., Chowdhury, S. M., & Hazra, S. (2018). Dynamics of the Sundarbans mangroves in Bangladesh under climate change. In R. J. Nicholls, C. W. Hutton, W. N. Adger,

S. E. Hanson, M. M. Rahman, & M. Salehin (Eds.), *Ecosystem services for well-being in deltas: Integrated assessment for policy analysis* (pp. 489–503). Cham: Springer. http://dx.doi.org/10.1007/978-3-319-71093-8_26.

Nicholls, R. J. (2018). Adapting to sea-level rise. In Z. Zommers & K. Alverson (Eds.), *Resilience: The science of adaptation to climate change* (pp. 14–29). Oxford, UK: Elsevier.

Nicholls, R. J., Brown, S., Goodwin, P., Wahl, T., Lowe, J., Solan, M., et al. (2018a). Stabilization of global temperature at 1.5°C and 2.0°C: Implications for coastal areas. *Philosophical Transactions of the Royal Society, 376*(2119). http://dx.doi.org/10.1098/rsta.2016.0448.

Nicholls, R. J., Hutton, C., Adger, W. N., Hanson, S. E., Rahman, M. M., & Salehin, M. (Eds.). (2018b). *Ecosystem services for well-being in deltas: Integrated assessment for policy analysis*. London, UK: Palgrave Macmillan.

Paola, C., Twilley, R. R., Edmonds, D. A., Kim, W., Mohrig, D., Parker, G., et al. (2010). Natural processes in delta restoration: Application to the Mississippi Delta. *Annual Review of Marine Science, 3*(1), 67–91. https://doi.org/10.1146/annurev-marine-120709-142856.

Paul, A. K., & Røskaft, E. (2013). Environmental degradation and loss of traditional agriculture as two causes of conflicts in shrimp farming in the southwestern coastal Bangladesh: Present status and probable solutions. *Ocean and Coastal Management, 85,* 19–28. https://doi.org/10.1016/j.ocecoaman.2013.08.015.

Payo, A., Mukhopadhyay, A., Hazra, S., Ghosh, T., Ghosh, S., Brown, S., et al. (2016). Projected changes in area of the Sundarban mangrove forest in Bangladesh due to SLR by 2100. *Climatic Change, 139*(2), 279–291. https://doi.org/10.1007/s10584-016-1769-z.

Quassem, M. A., Khan, B. U., Uddin, A. M. K., Ahmad, M., & Koudstaal, R. (2003). *A systems analysis of shrimp production* (Working Paper WP014). Program Development Office for Integrated Coastal Zone Management Plan (PDO-ICZMP).

Rahman, M. H., Lund, T., & Bryceson, I. (2011). Salinity impacts on agro-biodiversity in three coastal, rural villages of Bangladesh. *Ocean and Coastal Management, 54*(6), 455–468. https://doi.org/10.1016/j.ocecoaman.2011.03.003.

Ravallion, M. (2001). Growth, inequality and poverty: Looking beyond averages. *World Development, 29*(11), 1803–1815. https://doi.org/10.1016/S0305-750X(01)00072-9.

Renaud, F. G., Le, T. T. H., Lindener, C., Guong, V. T., & Sebesvari, Z. (2015). Resilience and shifts in agro-ecosystems facing increasing sea-level rise and salinity intrusion in Ben Tre Province, Mekong Delta. *Climatic Change, 133*(1), 69–84. https://doi.org/10.1007/s10584-014-1113-4.

Seijger, C., Douven, W., van Halsema, G., Hermans, L., Evers, J., Phi, H. L., et al. (2017). An analytical framework for strategic delta planning: Negotiating consent for long-term sustainable delta development. *Journal of Environmental Planning and Management, 60*(8), 1485–1509. https://doi.org/10.1080/09640568.2016.1231667.

Seijger, C., Ellen, G. J., Janssen, S., Verheijen, E., & Erkens, G. (2018). Sinking deltas: Trapped in a dual lock-in of technology and institutions. *Prometheus*, 1–21. http://dx.doi.org/10.1080/08109028.2018.1504867.

Smajgl, A., Toan, T. Q., Nhan, D. K., Ward, J., Trung, N. H., Tri, L. Q., et al. (2015). Responding to rising sea levels in the Mekong Delta. *Nature Climate Change, 5,* 167–174. https://doi.org/10.1038/nclimate2469.

Stive, M. J. F., Fresco, L. O., Kabat, P., Parmet, B. W. A. H., & Veerman, C. P. (2011). How the Dutch plan to stay dry over the next century. *Proceedings of the Institution of Civil Engineers—Civil Engineering, 164*(3), 114–121. https://doi.org/10.1680/cien.2011.164.3.114.

Suckall, N., Tompkins, E. L., Nicholls, R. J., Kebede, A. S., Lázár, A. N., Hutton, C., et al. (2018). A framework for identifying and selecting long term adaptation policy directions for deltas. *Science of the Total Environment, 633,* 946–957. https://doi.org/10.1016/j.scitotenv.2018.03.234.

Swapan, M. S. H., & Gavin, M. (2011). A desert in the delta: Participatory assessment of changing livelihoods induced by commercial shrimp farming in southwest Bangladesh. *Ocean and Coastal Management, 54*(1), 45–54. https://doi.org/10.1016/j.ocecoaman.2010.10.011.

Syvitski, J. P. M. (2008). Deltas at risk. *Sustainability Science, 3*(1), 23–32. https://doi.org/10.1007/s11625-008-0043-3.

Syvitski, J. P. M., Kettner, A. J., Overeem, I., Hutton, E. W. H., Hannon, M. T., Brakenridge, G. R., et al. (2009). Sinking deltas due to human activities. *Nature Geoscience, 2*(10), 681–686. https://doi.org/10.1038/ngeo629.

Szabo, S., Nicholls, R. J., Neumann, B., Renaud, F. G., Matthews, Z., Sebesvari, Z., et al. (2016). Making SDGs work for climate change hotspots. *Environment: Science and Policy for Sustainable Development, 58*(6), 24–33. http://dx.doi.org/10.1080/00139157.2016.1209016.

Tessler, Z. D., Vörösmarty, C. J., Overeem, I., & Syvitski, J. P. M. (2018). A model of water and sediment balance as determinants of relative sea level rise in contemporary and future deltas. *Geomorphology, 305,* 209–220. https://doi.org/10.1016/j.geomorph.2017.09.040.

United Nations. (2015). *Seventieth session agenda items 15 and 116 resolution adopted by the General Assembly on 25 September 2015*. Transforming Our World: The 2030 Agenda for Sustainable Development. New York, NY: UN General Assembly.

van Staveren, M. F., & van Tatenhove, J. P. M. (2016). Hydraulic engineering in the social-ecological delta: Understanding the interplay between social, ecological, and technological systems in the Dutch delta by means of "delta trajectories". *Ecology and Society, 21,* 1. https://doi.org/10.5751/ES-08168-210108.

van Wesenbeeck, B. K., Mulder, J. P. M., Marchand, M., Reed, D. J., de Vries, M. B., de Vriend, H. J., et al. (2014). Damming deltas: A practice of the past? Towards nature-based flood defenses. *Estuarine, Coastal and Shelf Science, 140*, 1–6. https://doi.org/10.1016/j.ecss.2013.12.031.

Welch, A. C., Nicholls, R. J., & Lázár, A. N. (2017). Evolving deltas: Co-evolution with engineered interventions. *Elementa Science of the Anthropocene, 5*, 49. http://dx.doi.org/10.1525/elementa.128.

Young, O. R., Berkhout, F., Gallopin, G. C., Janssen, M. A., Ostrom, E., & van der Leeuw, S. (2006). The globalization of socio-ecological systems: An agenda for scientific research. *Global Environmental Change, 16*(3), 304–316. https://doi.org/10.1016/j.gloenvcha.2006.03.004.

Index

R. J. Nicholls et al. (eds.), *Deltas in the Anthropocene*,
https://doi.org/10.1007/978-3-030-23517-8

www.ingramcontent.com/pod-product-compliance
Lightning Source LLC
LaVergne TN
LVHW082118200926
848484LV00008B/1972
* 9 7 8 1 0 1 3 2 7 5 0 1 2 *